A FARMER'S PRIMER ON GROWING SOYBEAN ON RICELAND

R.K. Pandey

International Rice Research Institute
and
International Institute of Tropical Agriculture

1987
International Rice Research Institute
Los Baños, Laguna, Philippines
P.O. Box 933, Manila, Philippines

The International Rice Research Institute (IRRI) was established in 1960 by the Ford and Rockefeller Foundations with the help and approval of the Government of the Philippines. Today IRRI is one of the 13 nonprofit international research and training centers supported by the Consultative Group on International Agricultural Research (CGIAR). The CGIAR is sponsored by the Food and Agriculture Organization (FAO) of the United Nations, the International Bank for Reconstruction and Development (World Bank), and the United Nations Development Programme (UNDP). The CGIAR consists of 50 donor countries, international and regional organizations, and private foundations.

IRRI receives support, through the CGIAR, from a number of donors including: the Asian Development Bank, the European Economic Community, the Ford Foundation, the International Development Research Centre, the International Fund for Agricultural Development, the OPEC Special Fund, the Rockefeller Foundation, the United Nations Development Programme, the World Bank, and the international aid agencies of the following governments: Australia, Belgium, Canada, China, Denmark, France, Federal Republic of Germany, India, Italy, Japan, Mexico, Netherlands, New Zealand, Norway, Philippines, Saudi Arabia, Spain, Sweden, Switzerland, United Kingdom, and United States.

The responsibility for this publication rests with the International Rice Research Institute.

Copyright © International Rice Research Institute 1987

All rights reserved. Except for quotations of short passages for the purpose of criticism and review, no part of this publication may be reproduced, stored in retrieval systems, or transmitted in any form or by any means, electronic, mechanical, photocopying, recording, or otherwise, without prior permission of IRRI. This permission will not be unreasonably withheld for use for noncommercial purposes. IRRI does not require payment for the noncommercial use of its published works, and hopes that this copyright declaration will not diminish the bona fide use of its research findings in agricultural research and development.

The designations employed and the presentation of the material in this publication do not imply the expression of any opinion whatsoever on the part of IRRI concerning the legal status of any country, territory, city, or area, or of its authorities, or concerning the delimitation of its frontiers or boundaries.

ISBN 971-104-168-5

Contents

The Soybean Crop 1
  The soybean crop 3
  The seed 15
  Seedling growth 23
  Growth stages — vegetative phase 35
  Growth stages — flowering 39
  Growth stages — pod development 45
  The roots 53
  Root nodules and nitrogen fixing 59
Growing Soybean 67
  Environment 69
  Water 77
  Choosing the right variety 83
  Tillage and planting 93
  Fertilizer and lime 103
  Growing conditions and dry matter production 113
  Harvesting and storing soybean 123
Increasing Yields and Profits 129
  Yield components 131
  Production factors 137
  Yield reducers — weeds 145
  Yield reducers — insect pests 163
  Yield reducers — diseases 177
Soybean in Other Cropping Systems 195
  Sequence cropping 197
  Intercropping 205
  Strip-cropping 213

Foreword

Soybean is a high-value crop in temperate zones where, with appropriate inputs, it is grown on a large scale. But soybean has been little exploited in the tropics because of constraints such as seed viability, free nodulation, and seed shattering. Other impediments are the lack of processing facilities and poor marketing structures.

Yet soybean has great potential — even for small farmers with limited resources — to fit into the rice-based cropping systems that dominate so much of the agricultural area in the tropics.

A soybean crop generates farm income in the off-season after the rice harvest. It enriches the soil and helps break the pest and disease cycle associated with continuous rice cropping. Nutritionally, soybean makes an excellent protein complement to the largely carbohydrate diets of farm families. Its unusually high oil content also puts soybean in demand both as a source of edible oil and as a raw material for the food and feed industries.

Soybean responds markedly — even dramatically — to its environment. To realize the full yield potential of soybean, farmers must know how the plant grows, what its critical growth stages are, and how to prevent stress at each stage. Although a large volume of literature is available on soybean farming in temperate zones, little has been published on growing soybean in the tropics. The International Institute of Tropical Agriculture (IITA) has recently developed soybean lines that combine seed longevity, free nodulation, and non-shattering with superior agronomic characters suitable for tropical agriculture.

A Farmer's Primer on Growing Soybean on Rice Land is intended specifically for farmers in the tropics whose productivity and income could be significantly increased by raising soybean.

Patterned after *A Farmer's Primer on Growing Rice*, which had been published in 33 languages by mid-1987, this Primer is designed for inexpensive copublication in developing countries. The English text has been blocked off from the line drawings. The International Rice Research Institute (IRRI) makes complimentary sets of the illustrations available to cooperators, who may translate, strip text onto the prints, and print translated editions on local presses.

This soybean Primer was made possible through a collaborative project sponsored by IRRI and IITA. A companion volume is *A Farmer's Primer on Growing Cowpea on Rice Land*.

Ms. Vrinda Kumble of Editorial Consultants Services, New Delhi, India, edited both the soybean and cowpea Primers. The illustrations were drawn by John Figarola, senior illustrator, IRRI Communication and Publications Department; and free-lance artists Joseph Figarola and Oscar Figuracion.

M.S. Swaminathan
Director General
International Rice Research
  Institute

Lawrence Stifel
Director General
International Institute
  of Tropical Agriculture

The soybean crop

The soybean crop

The soybean plant   5
Why grow soybean   6
Soybean enriches the soil   7
Breaks the pest and disease cycle   8
Adds to income   9
Soybean is a nutritious food   10
When to grow soybean   11
When to grow soybean   12
When to grow soybean   13
Duration of the crop   14

The soybean plant

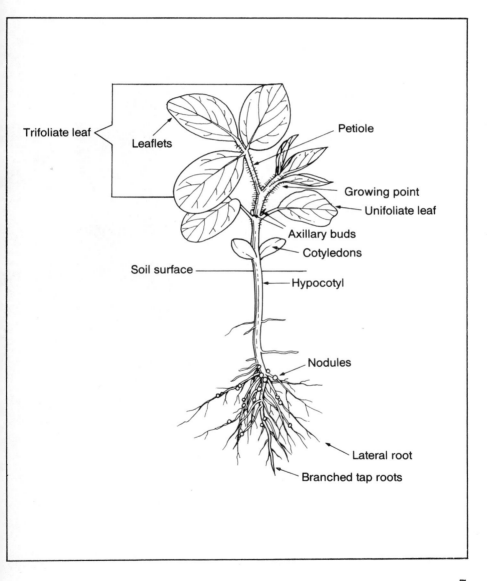

Why grow soybean

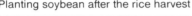
Planting soybean after the rice harvest

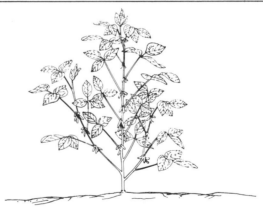

- Soybean is an annual legume crop that can be easily grown on riceland after the harvest of rice.
- With good management it can give high yields and profits.

Soybean enriches the soil

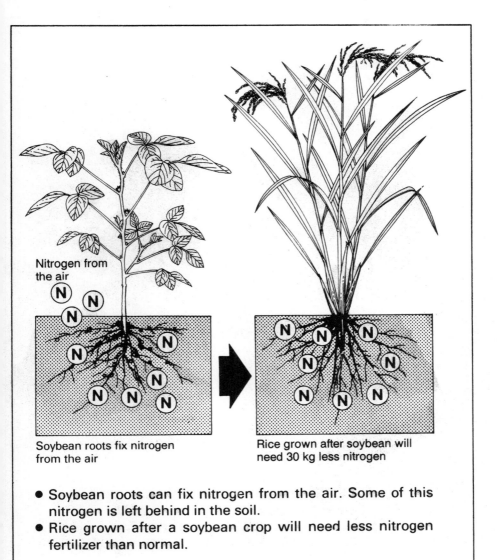

Soybean roots fix nitrogen from the air

Rice grown after soybean will need 30 kg less nitrogen

- Soybean roots can fix nitrogen from the air. Some of this nitrogen is left behind in the soil.
- Rice grown after a soybean crop will need less nitrogen fertilizer than normal.

Breaks the pest and disease cycle

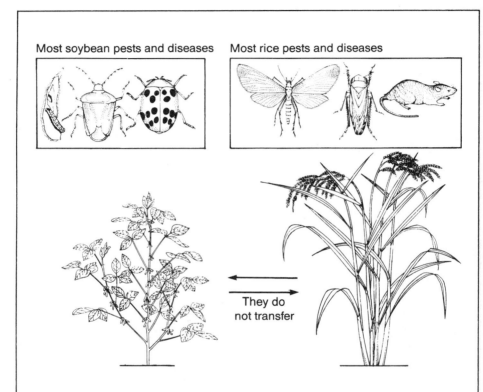

- Growing soybean in rotation with rice reduces pests and diseases on both crops because
  — most soybean pests and diseases do not transfer to rice;
  — most rice pests and diseases do not transfer to soybean.

Adds to income

Planting soybean after the rice harvest

Adds to income

- Growing soybean can provide jobs in the off season after the rice harvest and add to farm incomes.

Soybean is a nutritious food

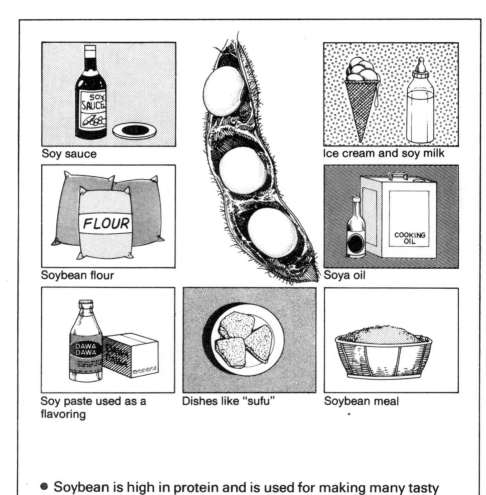

- Soy sauce
- Soybean flour
- Soy paste used as a flavoring
- Ice cream and soy milk
- Soya oil
- Dishes like "sufu"
- Soybean meal

- Soybean is high in protein and is used for making many tasty and wholesome foods.
- Oil from soybean can be used as a cooking oil. It also has many industrial uses.

When to grow soybean

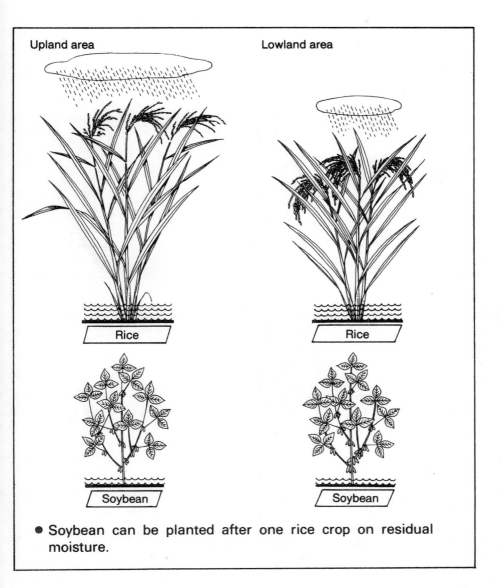

- Soybean can be planted after one rice crop on residual moisture.

When to grow soybean

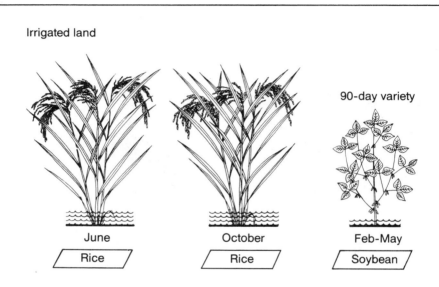

- Short-duration varieties can be planted after two rice crops on irrigated land.
- Yields are high when the crop is irrigated.

When to grow soybean

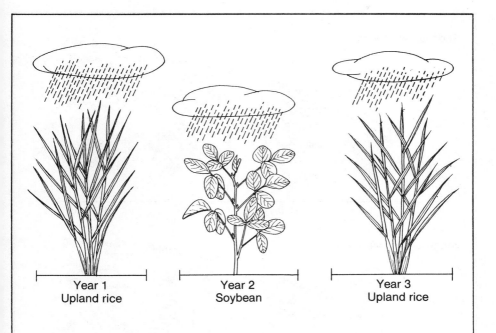

Year 1 Upland rice

Year 2 Soybean

Year 3 Upland rice

- If soil is well drained, soybean can be grown in alternate years in the rainy season, in rotation with upland rice.
- This system maintains soil fertility.

Duration of the crop

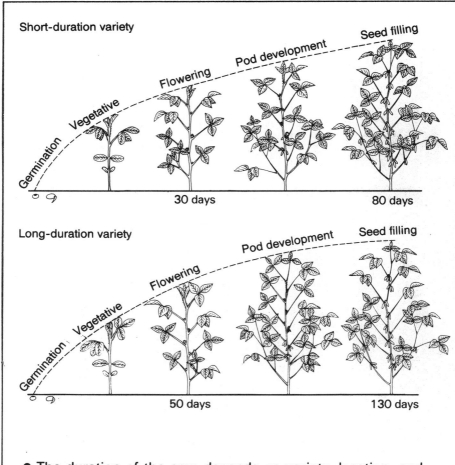

- The duration of the crop depends on variety, location, and season.
- Sowing to harvest may take 80 days from short-duration varieties; 130 days for long-duration ones.

The seed

The soybean seed   17
Parts of the seed   18
Germination   19
Conditions needed for germination — water   20
Conditions needed for germination — air and warmth   21
Conditions needed for germination — seed quality   22

The soybean seed

Seeds vary in size, shape, and color

Small seeded
(10 g per 100 seeds)

Large seeded
(25 g per 100 seeds)

Black

Brown

Yellow

Cream

Greenish

- Soybean seeds vary in size and shape.
- Color may be white, cream, yellow, green, brown, or black.

Parts of the seed

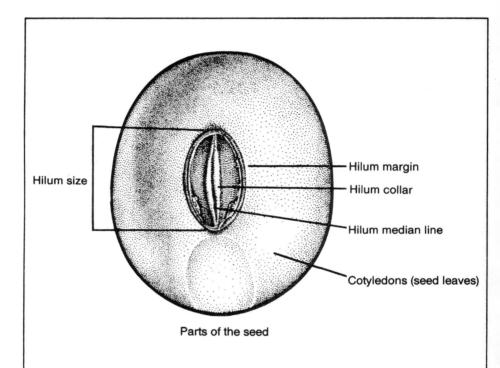

Parts of the seed

- The seed leaves contain food for the growing embryo: about 40 percent protein, 18 percent fat, and the rest starch and sugar.

Germination

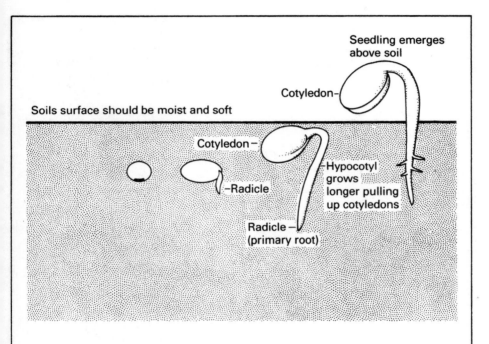

- The soybean seed begins germinating by absorbing about half its weight of water.
- The radicle or primary root is the first to grow from the seed.

Conditions needed for germination — water

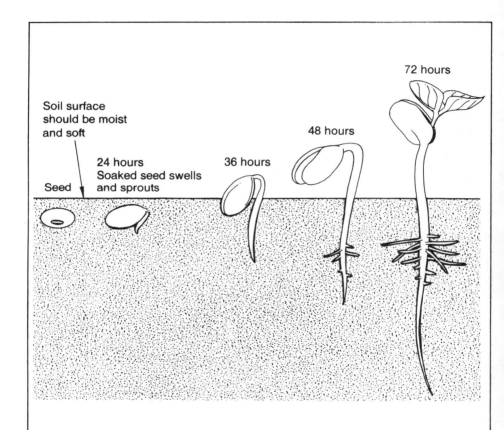

- Water is the first need of a seed for germination.
- Many activities go on inside the germinating seed. Starch, proteins, and fats stored in the seed are changed into simple foods for the growing embryo.

Conditions needed for germination — air and warmth

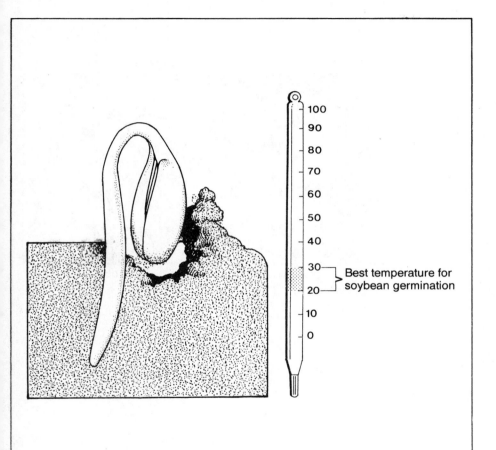

Best temperature for soybean germination

- The germinating soybean seed needs oxygen from the air.
- If the seed is planted too deep the embryo gets no air and cannot grow.
- The best temperature for germination is 20 to 30°C. Too high or low a temperature reduces germination.

Conditions needed for germination — seed quality

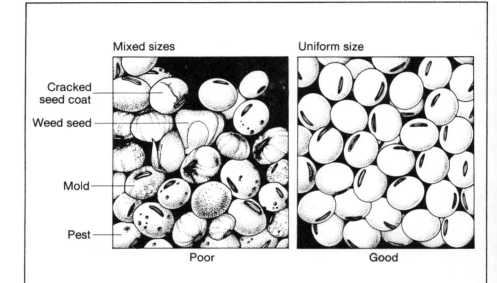

- For good germination and plant stand, seed planted should be clean and free of pests and diseases.
- Seed for planting should be stored no more than 4 months unless it is kept in cold storage.

Seedling growth

Seedling growth **25**
Factors affecting seedling growth — water **26**
Factors affecting seedling growth — temperature **27**
Factors affecting seedling growth — light intensity **28**
Factors affecting seedling growth — nutrients **29**
Factors affecting seedling growth — plant density **30**
Factors affecting seedling growth — weeds, insect pests, and diseases **31**

Seedling growth

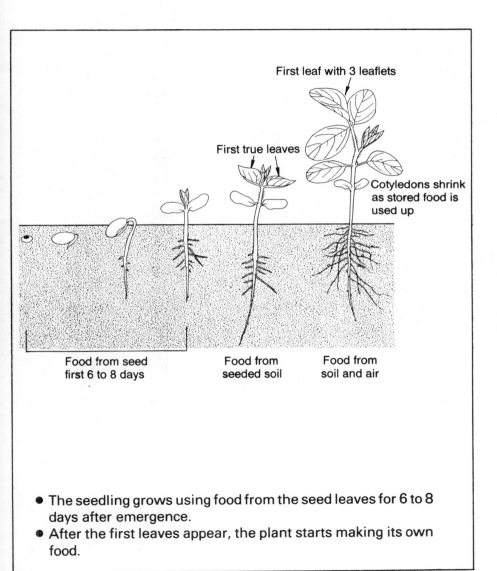

- The seedling grows using food from the seed leaves for 6 to 8 days after emergence.
- After the first leaves appear, the plant starts making its own food.

Factors affecting seedling growth — water

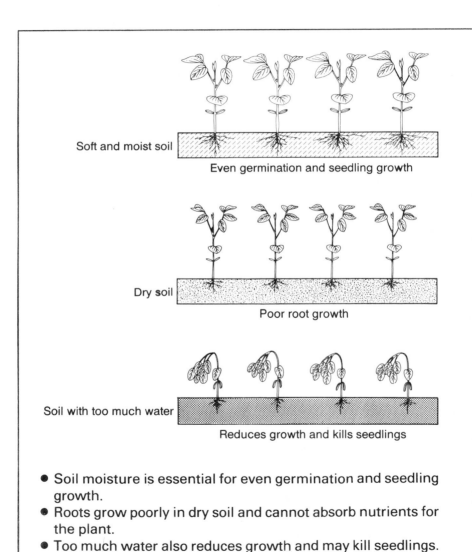

- Soil moisture is essential for even germination and seedling growth.
- Roots grow poorly in dry soil and cannot absorb nutrients for the plant.
- Too much water also reduces growth and may kill seedlings.

Factors affecting seedling growth — temperature

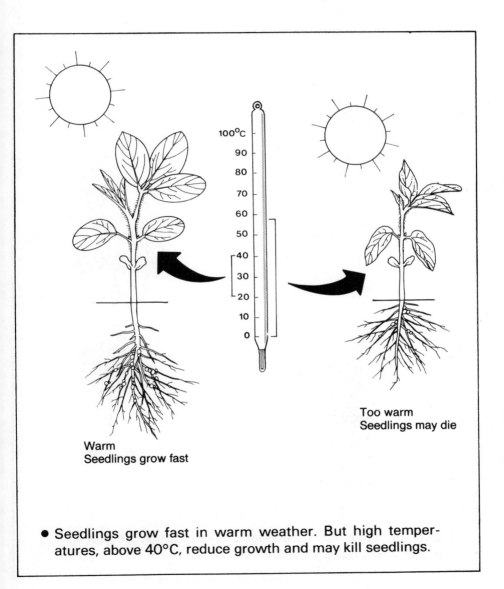

Warm
Seedlings grow fast

Too warm
Seedlings may die

- Seedlings grow fast in warm weather. But high temperatures, above 40°C, reduce growth and may kill seedlings.

Factors affecting seedling growth — light intensity

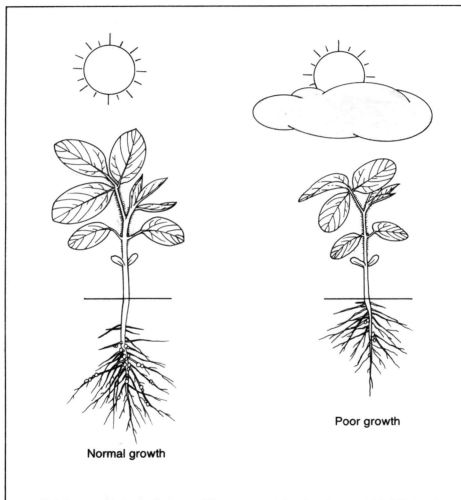

Normal growth

Poor growth

- Bright sunlight helps seedlings grow vigorously. Lack of light makes seedlings pale and weak-stemmed.

Factors affecting seedling growth — nutrients

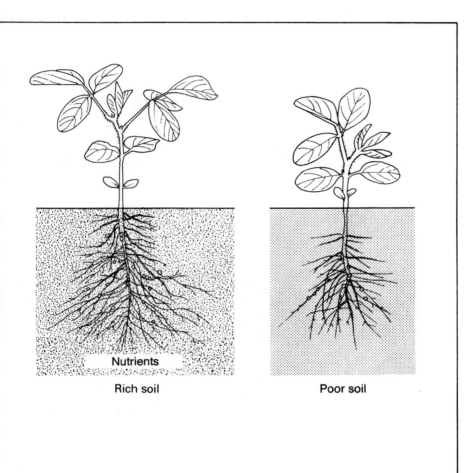

Rich soil   Poor soil

- To grow fast, seedlings need readily available nutrients. In poor soils, fertilizer may be needed at planting to start rapid growth.

Factors affecting seedling growth — plant density

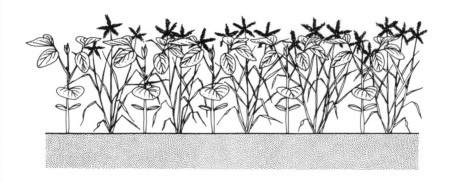

- Seedlings growing too close together grow too tall and lodge easily.
- Seedlings growing too far apart allow too much weed growth.

Factors affecting seedling growth — weeds, insect pests, and diseases

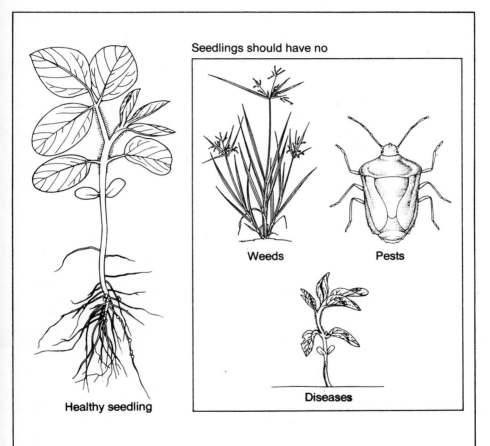

- Weeds rob seedlings of nutrients.
- Insect pests eat young leaves and stems and may kill seedlings.
- Soil-borne diseases stunt or kill young seedlings.

Growth stages

Growth stages of soybean — vegetative phase **35**
Growth stages of soybean — vegetative phase **36**
Growth stages of soybean — late vegetative phase **37**
Branching **38**

Growth stages of soybean — vegetative phase

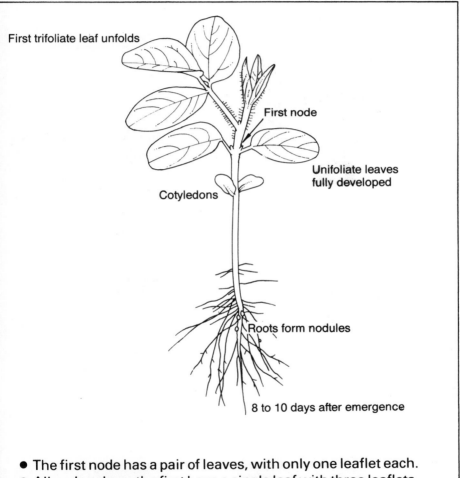

- The first node has a pair of leaves, with only one leaflet each.
- All nodes above the first have a single leaf with three leaflets.
- Roots start forming nodules about one week after the seedling emerges above the soil.

Growth stages of soybean — vegetative phase

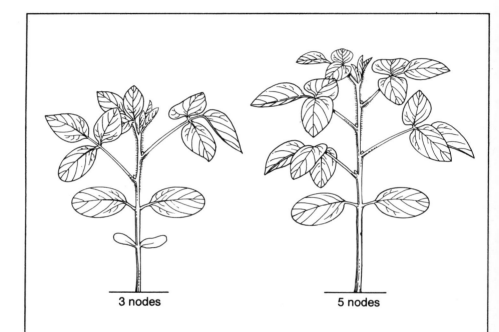

3 nodes

5 nodes

- The stem grows rapidly, with a new leaf unrolling at each node.
- Roots begin actively fixing nitrogen by the time the second or third node has developed.

Growth stages — late vegetative phase

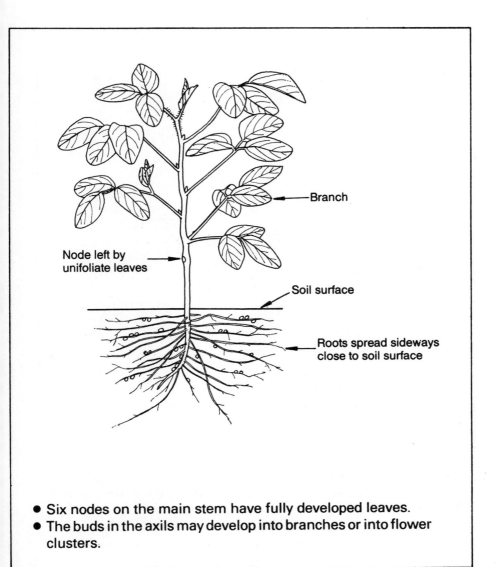

- Six nodes on the main stem have fully developed leaves.
- The buds in the axils may develop into branches or into flower clusters.

Branching

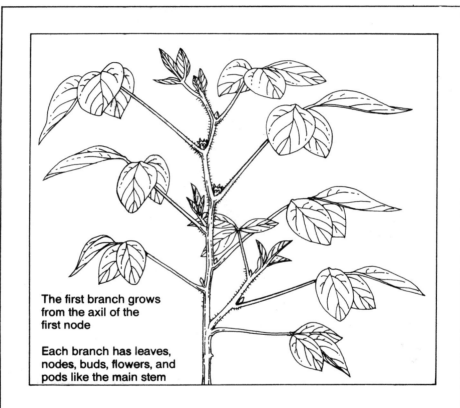

The first branch grows from the axil of the first node

Each branch has leaves, nodes, buds, flowers, and pods like the main stem

- Branching starts when the plant is about 20 cm tall. The number of branches depends on the soybean variety and plant density.
- Branches are useful in making up some yield where plant density is low, or when the main stem tip is damaged.

Growth stages — flowering

Flowering 41
Flowering 42
Flowering pattern —determinate varieties 43
Flowering pattern —indeterminate varieties 44

Flowering

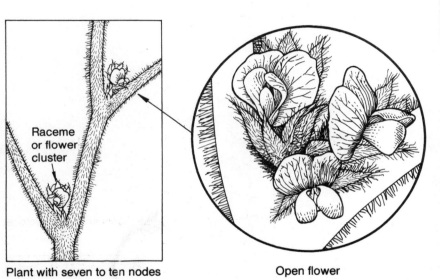

Plant with seven to ten nodes

Open flower

- Soybean flowers grow in clusters called racemes.
- Number of days to first flower depends on soybean variety, daylength, and temperature.

Flowering

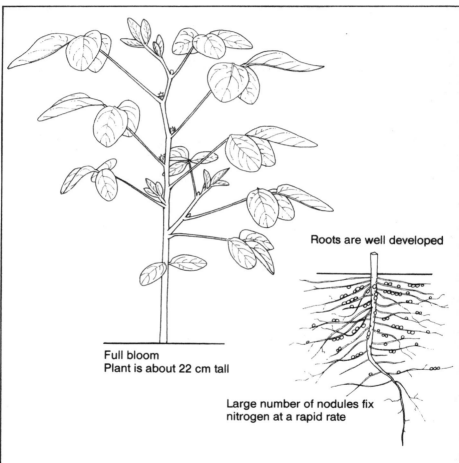

Roots are well developed

Full bloom
Plant is about 22 cm tall

Large number of nodules fix nitrogen at a rapid rate

- By full bloom stage the plant has accumulated 25 to 30 percent of its total dry weight. From now on dry weight increases rapidly.
- Nitrogen fixing is also very rapid at this stage.

Flowering pattern — determinate varieties

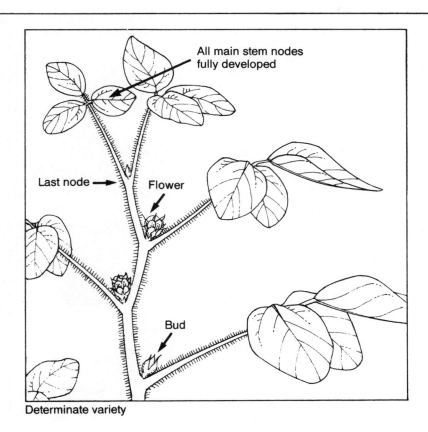
Determinate variety

- The flowering pattern of the soybean plant depends on the variety.
- Determinate varieties begin flowering when most of the nodes on the main stem have developed. Flowering starts at the upper nodes and progresses downwards and upwards from there.

Flowering pattern — indeterminate varieties

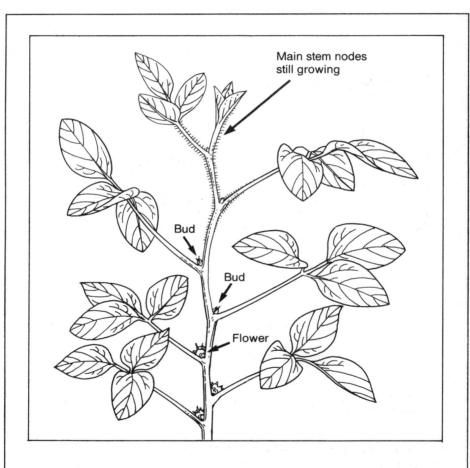

- Indeterminate varieties begin flowering when less than half the nodes on the main stem have developed.
- Flowering starts at the lower nodes, which develop pods while upper nodes are still flowering.

Growth stages —pod development

Pod formation **47**
Full pod **48**
Seed filling **49**
Seed filling **50**
Ripening **51**
Full maturity **52**

Pod formation

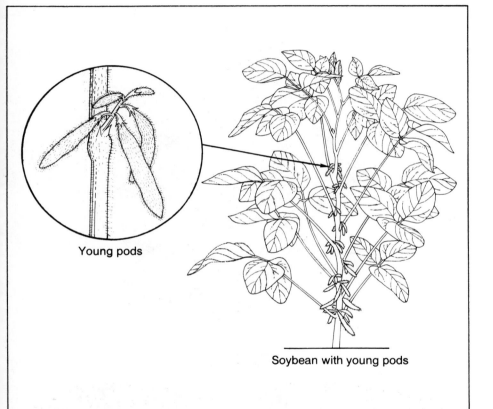

Young pods

Soybean with young pods

- Only about 40 percent of all the flowers on a plant develop into pods.
- High temperatures (above 35°C), lack of water, and lack of nutrients at this stage can cause young pods also to drop.

Full pod

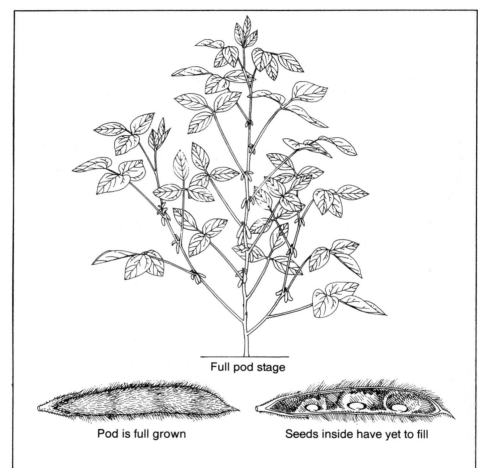

Full pod stage

Pod is full grown   Seeds inside have yet to fill

- Pods grow rapidly to their full length and width.
- Full-pod stage is the most sensitive to stress. Lack of water or nutrients or very high temperatures now will reduce yields drastically.

Seed filling

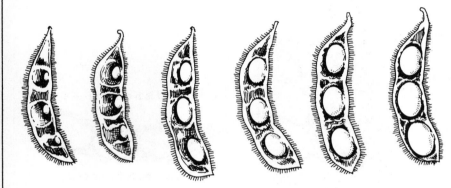

Seeds fill rapidly

- Seed yields depend upon the rate and length of time that dry weight accumulates in the seed.
- Nitrogen-fixing rate is highest at the beginning of this stage but drops sharply later.

Seed filling

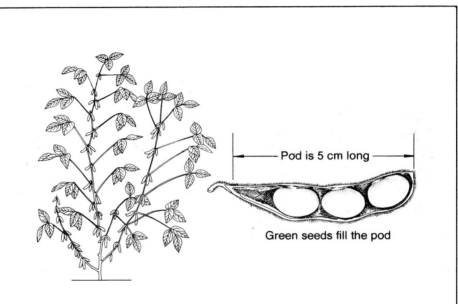

- Nutrients accumulated in the leaves and other vegetative parts are transferred to the seeds.
- Leaves begin to yellow and oldest leaves start falling.

Ripening

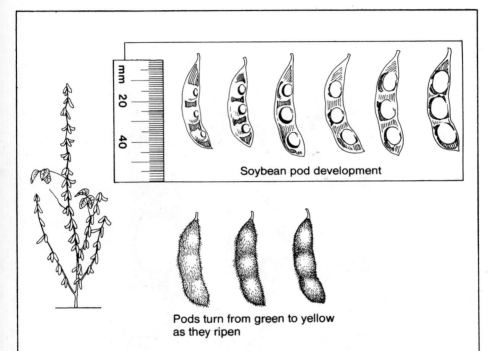

Soybean pod development

Pods turn from green to yellow as they ripen

- As pods and seeds develop, they are less likely to drop off. Number of pods per plant is set.
- Soybean seeds turn yellow as they ripen. They must dry before harvest.

Full maturity

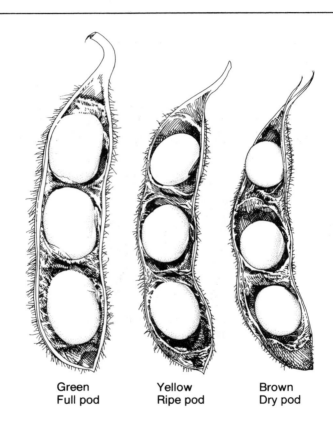

Green
Full pod

Yellow
Ripe pod

Brown
Dry pod

- After 95 percent of the pods have turned yellow, 5 to 7 days of drying weather are needed. Rain at this time can spoil the seeds.
- Timely harvesting is necessary to prevent seed loss in the field.

The roots

Functions of the roots **55**
Root development **56**
Root distribution **57**

Functions of the roots

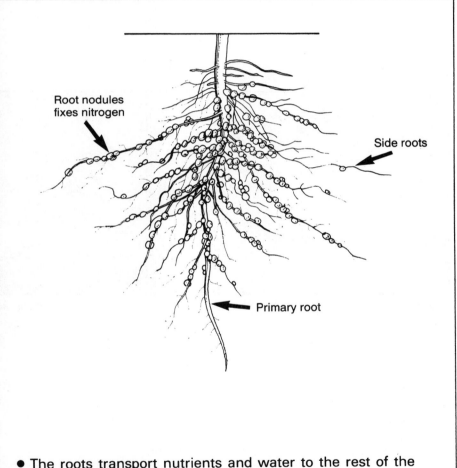

- The roots transport nutrients and water to the rest of the plant.
- They support the shoot and its parts.
- In soybean, the roots also fix nitrogen.

Root development

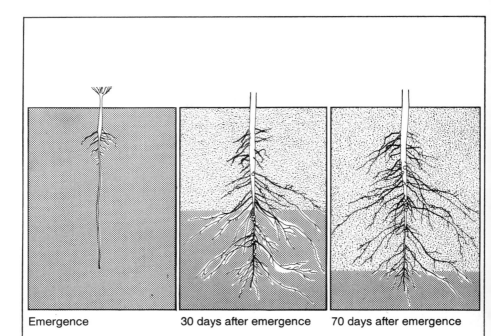

Emergence   30 days after emergence   70 days after emergence

- Roots develop much faster than the shoot.
- The side roots spread horizontally close to the soil surface for several weeks early in the season.
- As the soil moisture dries out, roots grow deep into the soil to absorb water and nutrients.

Root distribution

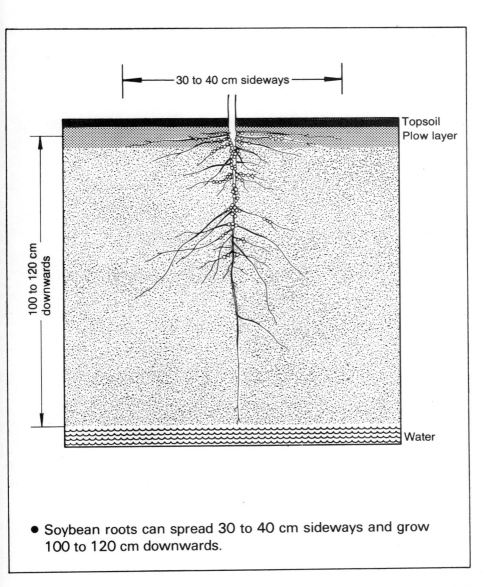

- Soybean roots can spread 30 to 40 cm sideways and grow 100 to 120 cm downwards.

Root nodules and nitrogen fixing

Root nodules **61**
Conditions affecting nodule growth and nitrogen fixing **62**
Conditions affecting nitrogen fixing — soil nitrogen and phosphorus **63**
Conditions affecting nitrogen fixing — temperature and daylength **64**
Conditions affecting nitrogen fixing — soil rhizobia **65**

Root nodules

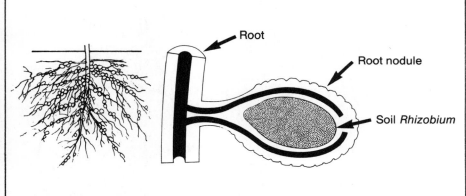

Nodulated roots   A root nodule looks like this inside

- Nodules are small lumps that form on soybean roots. Soil bacteria, known as *Rhizobium japonicum,* live inside the nodules.
- The bacteria fix nitrogen from air into forms that the plant can use.
- Nitrogen fixing increases as the plant grows, reaching a peak when seed filling begins.

Conditions affecting nodule growth and nitrogen fixing

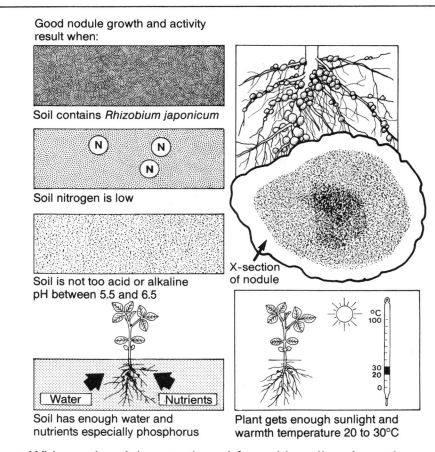

Good nodule growth and activity result when:

Soil contains *Rhizobium japonicum*

Soil nitrogen is low

Soil is not too acid or alkaline pH between 5.5 and 6.5

X-section of nodule

Soil has enough water and nutrients especially phosphorus

Plant gets enough sunlight and warmth temperature 20 to 30°C

- With good nodule growth and favorable soil and weather conditions, a soybean crop can fix up to 280 kg of nitrogen per hectare over the whole season.
- A healthy nodule is pink or red on the inside. White, brown, or green nodules mean that nitrogen is not being fixed.

Conditions affecting nitrogen fixing — soil nitrogen and phosphorus

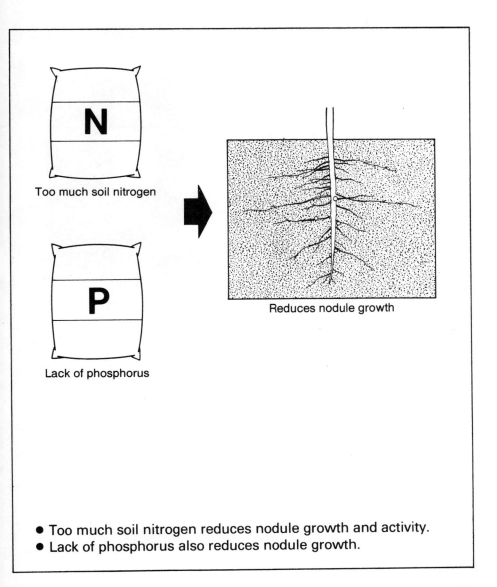

Too much soil nitrogen

Lack of phosphorus

Reduces nodule growth

- Too much soil nitrogen reduces nodule growth and activity.
- Lack of phosphorus also reduces nodule growth.

Conditions affecting nitrogen fixing — temperature and daylength

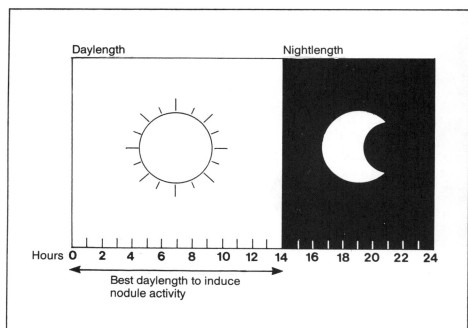

- Warm days and cool nights increase nodule activity.
- Daylength should be about 10 to 14 hours.

Conditions affecting nitrogen fixing — soil rhizobia

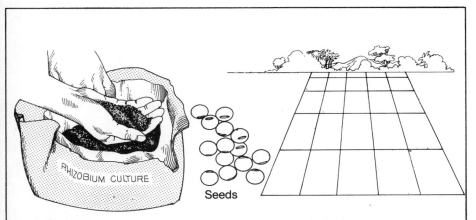

Treat seed with
Rhizobium japonicum culture

1 pocket (400 g) culture is enough for half hectare (30 to 40 kg seed)

- Soybean needs the right kind of soil bacteria to grow root nodules.
- Seeds should be treated with *Rhizobium* culture before planting every year.

Growing soybean

Growing soybean — environment

Temperature and rainfall **71**
Daylength **72**
Light intensity **73**
Soil **74**
Soil pH **75**

Temperature and rainfall

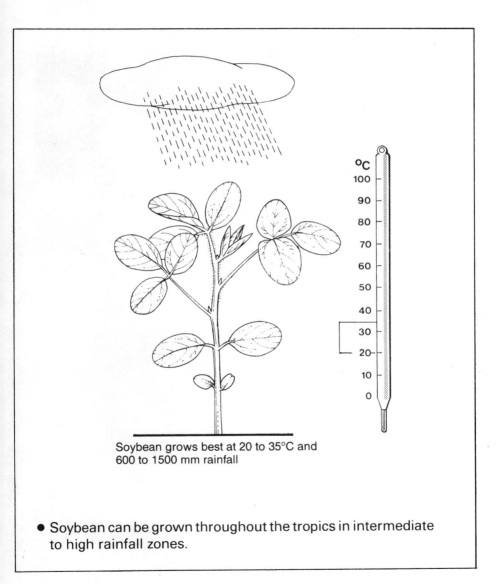

Soybean grows best at 20 to 35°C and 600 to 1500 mm rainfall

- Soybean can be grown throughout the tropics in intermediate to high rainfall zones.

Daylength

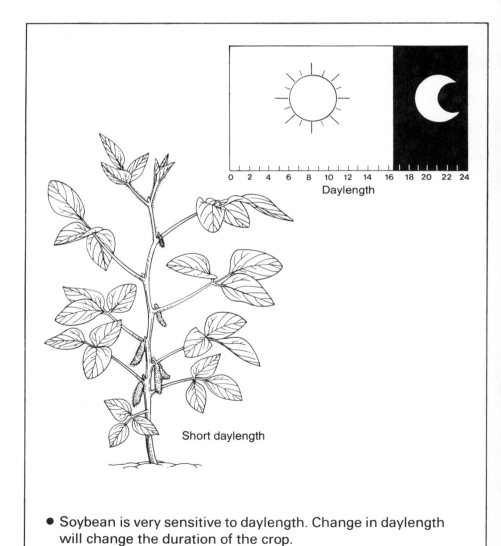

Short daylength

- Soybean is very sensitive to daylength. Change in daylength will change the duration of the crop.

Light intensity

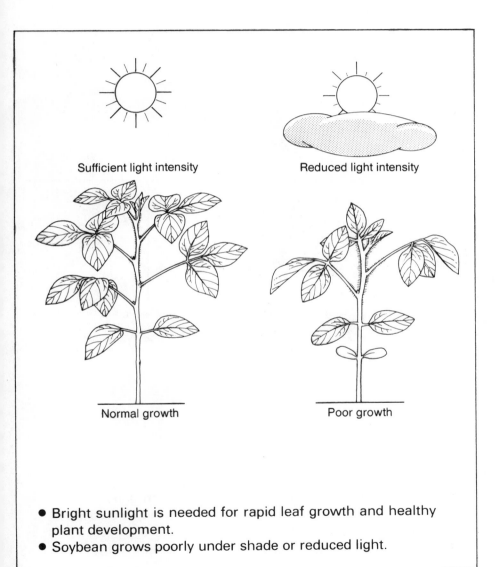

Soil

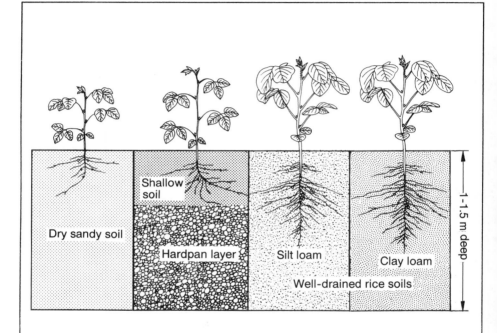

- Soybean needs deep, well-drained soils with good water-holding capacity. It cannot be grown on sandy dry soils or shallow soils over hardpan.
- Well-drained rice soils are suited to soybean growing.

Soil pH

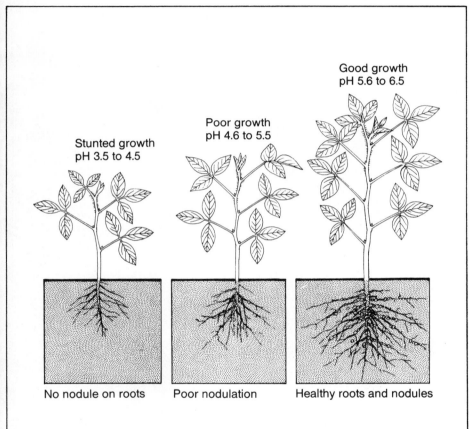

- Acid soils are not suited to growing soybean.
- Soil pH should be between 5.5 and 6.5.
- Adding lime will improve acid soils enough to grow a good soybean crop.

Growing soybean — water

Water needs **79**
When water is most needed **80**
How much water **81**
Irrigating soybean **82**

Water needs

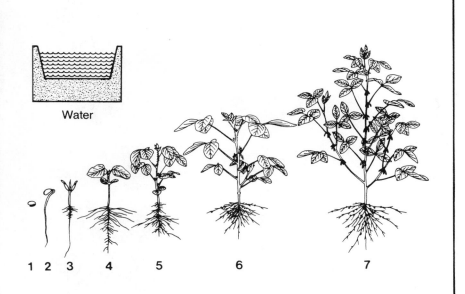

- The soybean crop needs about 400 to 550 mm available water over the whole growing season.
- Much more or much less than this will reduce yields.

When water is most needed

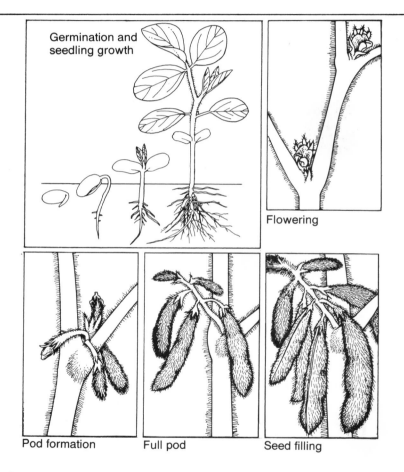

- Soil moisture is most needed during germination and early seedling growth and from pod formation through seed filling.
- Lack of water at these critical stages will drastically reduce yields.

How much water

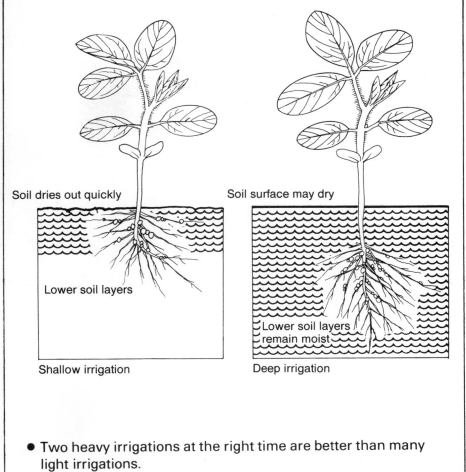

- Two heavy irrigations at the right time are better than many light irrigations.
- Lower soil layers remain moist even when soil surface dries out. Roots grow deep to absorb nutrients.
- Waterlogging is bad for the crop.

Irrigating soybean

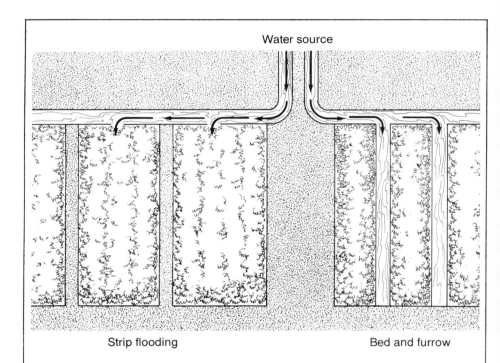

Strip flooding   Bed and furrow

- Strip flooding only needs a few irrigation channels. But it waters unevenly and may waste water.
- Preparing a bed-and-furrow system is costly at first. But it applies water evenly and does not waste water.

Growing season — choosing the right variety

Choosing the right variety **85**
The right variety — duration **86**
The right variety — pest and disease resistance **87**
The right variety — drought tolerance **88**
The right variety — lodging resistance **89**
The right variety — shattering resistance **90**
The right variety — free nodulation **91**

Choosing the right variety

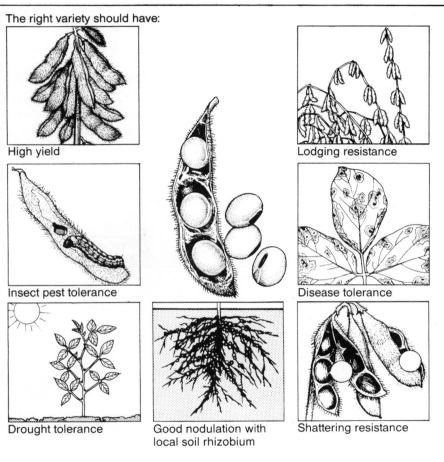

The right variety should have:

High yield

Lodging resistance

Insect pest tolerance

Disease tolerance

Drought tolerance

Good nodulation with local soil rhizobium

Shattering resistance

- Soybean yields depend on variety and growing conditions.
- Choose the variety to fit the cropping system and available water.
- Plant high-yielding varieties.

The right variety — duration

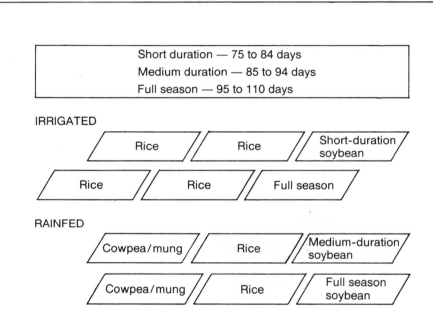

- Full-season varieties usually yield more than short-duration ones.
- Short-duration varieties allow more than one crop to be grown in sequence.
- Intermediate varieties yield well under most growing conditions.

The right variety — pest and disease resistance

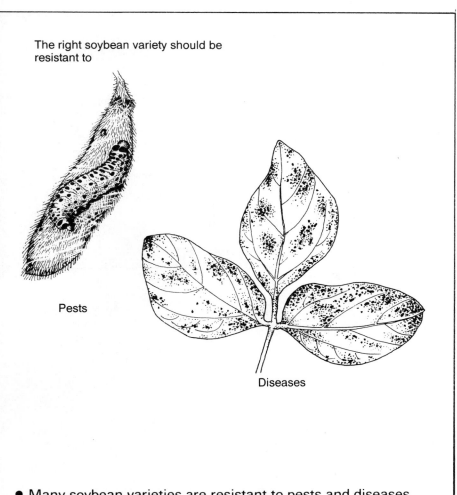

The right soybean variety should be resistant to

Pests

Diseases

- Many soybean varieties are resistant to pests and diseases. Choose the variety resistant to the most damaging pests and diseases in your area.

The right variety — drought tolerance

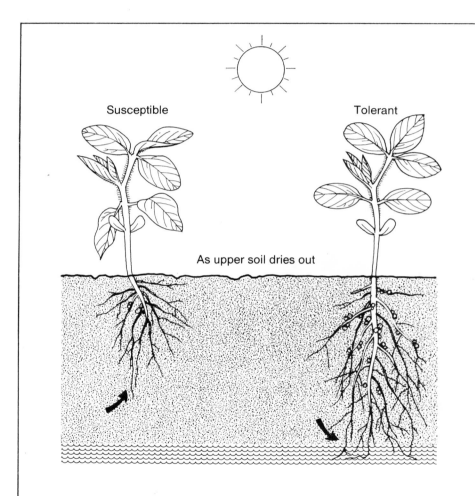

- Lack of water can reduce crop yields by 20 to 60 percent.
- In rainfed areas grow deep-rooted varieties of soybean that can withstand drought and draw on subsoil water.

The right variety — lodging resistance

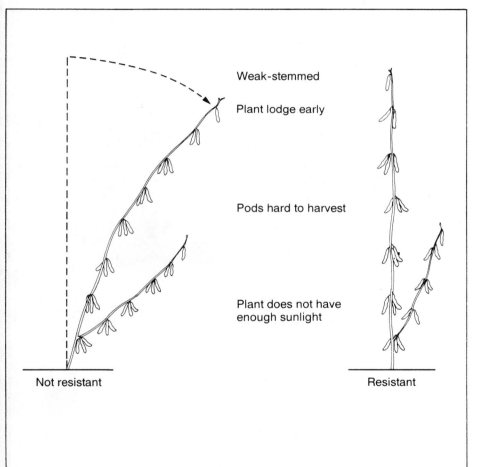

- Lodging, or falling over of plants, is common in irrigated soybean or in the rainy season.
- Lodging reduces yields by 5 to 30 percent. Plant lodging-resistant varieties.

The right variety — shattering resistance

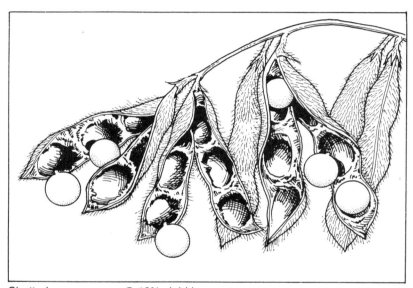

Shattering may cause 5-10% yield loss

- Soybean pods break open easily, and seeds are lost, if harvest is delayed.
- Plant shatter-resistant varieties.

The right variety — free nodulation

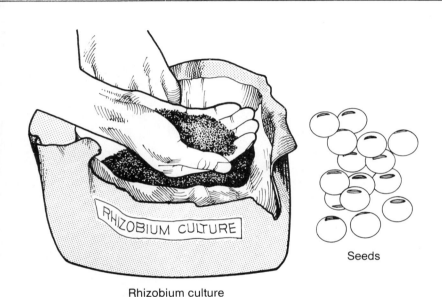

Rhizobium culture

Seeds

- Some soybean varieties can grow nodules with local soil rhizobia.
- In fields where soybean has not been grown for more than 5 years, plant free-nodulating varieties.
- Seed treatment with *Rhizobium* culture improves nodulation in all varieties.

Tillage and planting

Preparing the land — high tillage **95**
Preparing the land — zero tillage **96**
Planting season and date **97**
Plant density **98**
Row spacing **99**
Planting method **100**
Planting depth **101**

Preparing the land — high tillage

Deep plowing and two cross harrowings

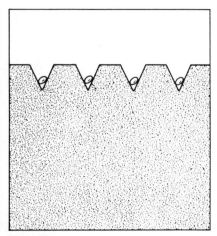

High tillage

- High tillage is common in irrigated areas where water is easily available. High tillage
  — airs the soil
  — helps seeds germinate and roots grow deep
  — controls weeds.
- But high tillage
  — is costly
  — delays planting
  — dries out the soil.

Preparing the land — zero tillage

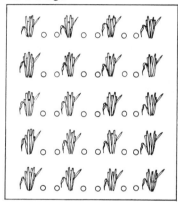

Zero tillage

No plowing
No harrowing

Soybean seed is planted in a furrow or dibbled at the base of rice stubble

- Zero tillage is common in rainfed areas, especially after lowland rice. Zero tillage
    - saves labor and costs
    - allows planting at once
    - makes full use of soil moisture.
- But zero tillage
    - does not air the soil
    - does not help roots grow deep
    - lets weeds grow.

Planting season and date

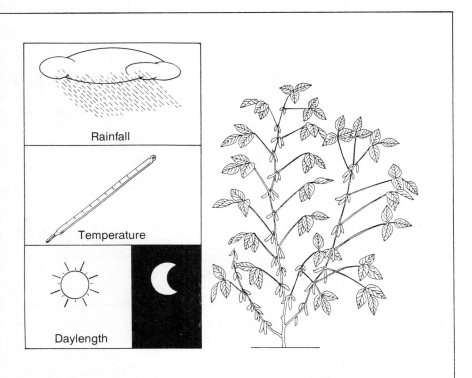

- The season for planting soybean depends on rainfall, temperature, and daylength.
- The best planting date differs with season and location. Short winter days usually lower seed yields.

Plant density

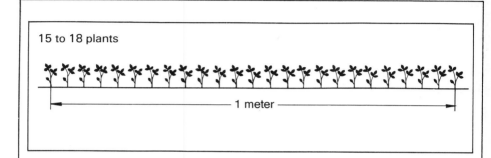

15 to 18 plants

1 meter

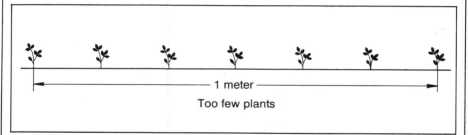

1 meter

Too few plants

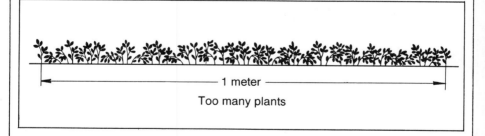

1 meter

Too many plants

- Best seed yields are obtained with 15 to 18 plants per meter row (planting 60 to 80 kg seed per hectare).
- Too dense a planting increases lodging.

Row spacing

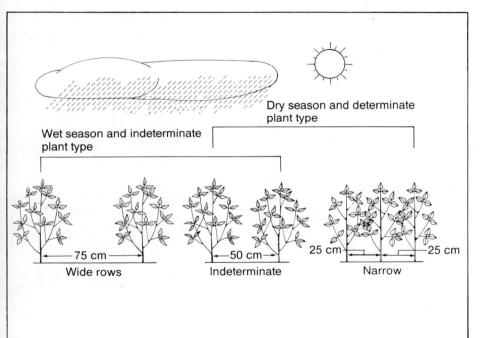

- Space between rows varies with plant type and season.
- Narrow spacing usually gives higher yields than wide row spacing.

Planting method

Dibble seed at the base of rice stubble after rice harvest

- Drill seed in rows by hand or animal-drawn seeder.
- Dibble seed at the base of rice stubble after rice harvest.

Planting depth

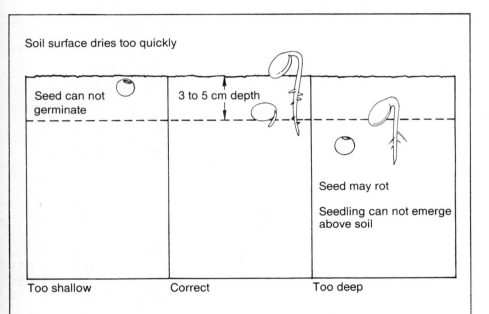

- The best depth for planting soybean seed is 3 to 5 cm.
- Planting deeper than 5 cm delays emergence. Seed may rot and plant stands will be uneven.

Fertilizer and lime

Why apply fertilizer **105**
Yield increases from fertilizer applied **106**
Organic fertilizer **107**
Fertilizer — nitrogen **108**
Fertilizer — phosphorus **109**
Fertilizer — potassium **110**
Fertilizer — micronutrients **111**
Lime **112**

Why apply fertilizer

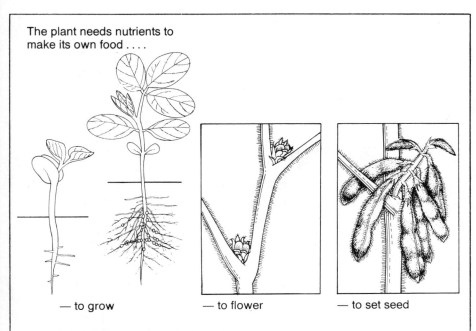

The plant needs nutrients to make its own food . . . .

— to grow — to flower — to set seed

Nutrients that the soybean plant needs

Nitrogen	Magnesium	Molybdenum	Zinc
Phosphorus	Sulfur	Boron	Manganese
Potassium	Calcium	Iron	

- A soybean plant needs many nutrients for healthy growth and high yields. Many of these are supplied by the soil.
- Where soils are poor, these nutrients must be supplied by adding fertilizer.

Yield increases from fertilizer applied

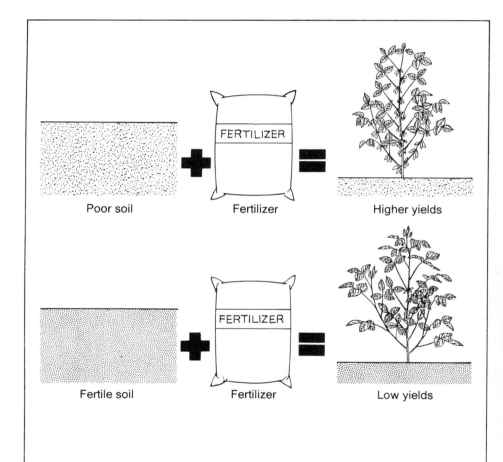

Poor soil + Fertilizer = Higher yields

Fertile soil + Fertilizer = Low yields

- Yield increases from fertilizer will be highest on poor soils.
- Do not apply fertilizer to fertile soil. It will give too much leafy growth and reduce yields.

Organic fertilizer

- Add organic fertilizer in any amount possible. Very large amounts are needed to improve seed yields significantly. But even small amounts will improve the soil structure and plant growth.

Fertilizer — nitrogen

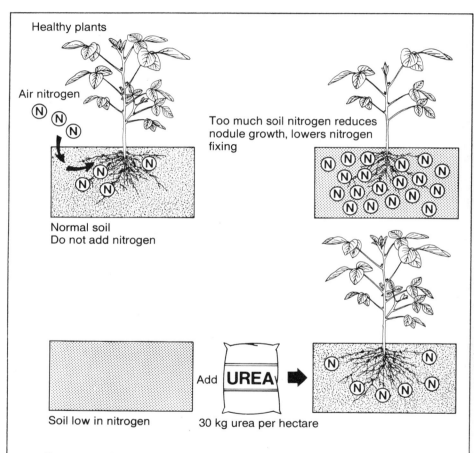

- On normal soils, soybean needs no added nitrogen fertilizer, because its roots can change nitrogen from the air into forms that the plant can use.
- But on poor soils, apply 30 kg urea per hectare at planting, to start the crop.

Fertilizer — phosphorus

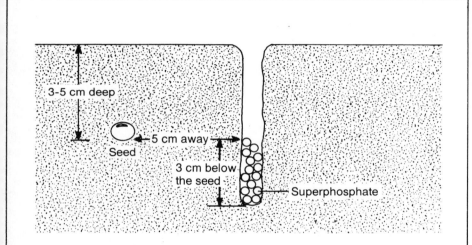

- Soybean needs phosphorus for root and nodule growth and for flowering.
- If soil is low in phosphorus, add 180 kg single superphospate at planting time.

Fertilizer — potassium

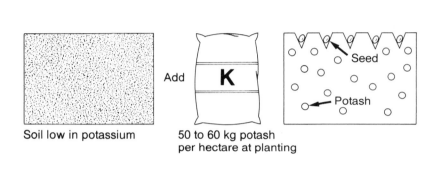

Soil low in potassium   50 to 60 kg potash per hectare at planting

- Most soils have enough potassium for the soybean crop.
- Where soil tests low in potassium, 50 to 60 kg potash per hectare should be applied at planting.

Fertilizer — micronutrients

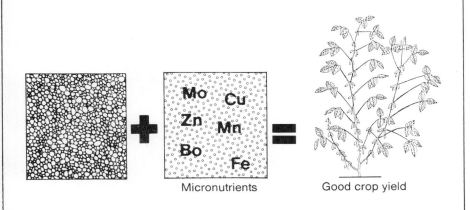

Micronutrients

Good crop yield

- Micronutrients are needed only in very small amounts, and most soils contain enough.
- A micronutrient should be applied only where soil tests show a lack of it.

Lime

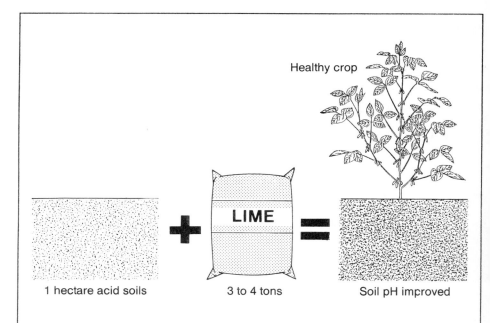

- To grow a soybean crop on acid soils, 3 to 4 tons of lime per hectare must be applied.
- Liming is costly but effectively improves soil pH.

Growing conditions and dry matter production

Dry matter production   **115**
Dry matter distribution   **116**
Factors affecting dry matter production   **117**
Factors affecting dry matter production — leaf growth   **118**
Factors affecting dry matter production — sunlight   **119**
Factors affecting dry matter production — water   **120**
Factors affecting dry matter production — nutrients   **121**

Dry matter production

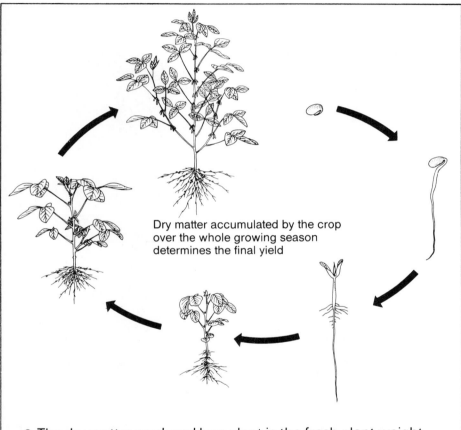

Dry matter accumulated by the crop over the whole growing season determines the final yield

- The dry matter produced by a plant is the fresh plant weight minus water.
- The dry matter accumulated by the crop over the whole growing season determines the final yield.
- Growing conditions at each stage affect dry matter accumulated.

Dry matter distribution

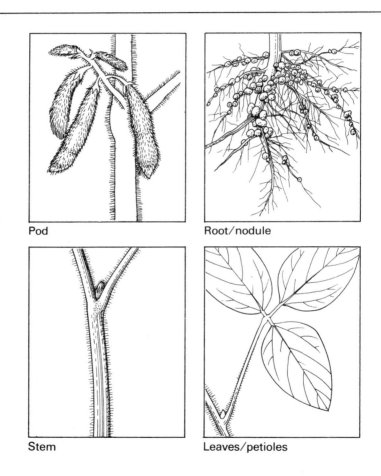

Pod

Root/nodule

Stem

Leaves/petioles

- High seed yield depends on proper distribution of dry matter to roots, stem, leaves, and pods.

Factors affecting dry matter production

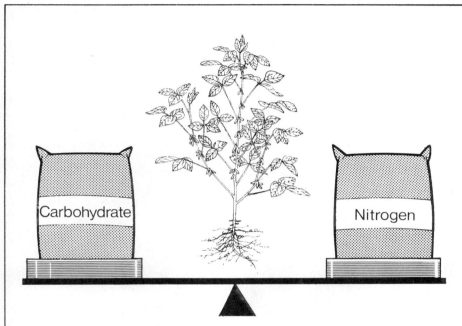

A balance of carbohydrate and nitrogen

- For best dry matter production, there should be a balance between the carbohydrate made by the leaves and the nitrogen fixed by the roots.

Factors affecting dry matter production — leaf growth

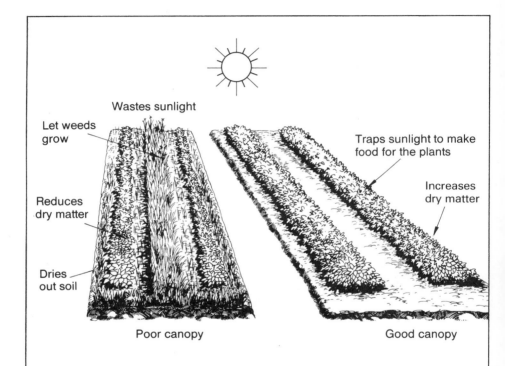

- Rapid leaf growth traps sunlight to make carbohydrate for the plant.
- The upper leaves should form an umbrella, or canopy, shading the ground between plant rows. Some sunlight should reach the lower leaves.

Factors affecting dry matter production — sunlight

Light increases dry matter production

Shade reduces dry matter production

- Bright sunlight increases dry matter produced.
- When soybean is grown in the shade, dry matter is reduced as shade increases.

Factors affecting dry matter production — water

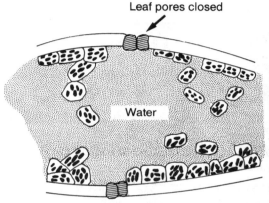

Cross section of a highly magnified leaf

Leaf pores closed

Water

With little water, leaf pores close, reducing food made by leaves

Soil nutrient

With too much water, roots can not absorb soil nutrients

- The maximum dry matter is produced when the soil contains the right amount of moisture.
- Soybean should be grown on deep, well-drained soils.

Factors affecting dry matter production — nutrients

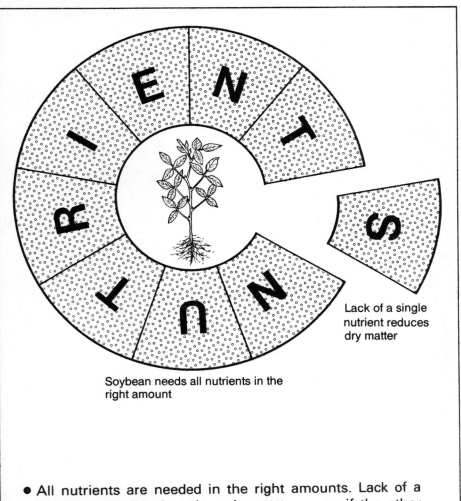

Soybean needs all nutrients in the right amount

Lack of a single nutrient reduces dry matter

- All nutrients are needed in the right amounts. Lack of a nutrient will sharply reduce dry matter, even if the other nutrients are well supplied.

Harvesting and storing soybean

Harvesting **125**
Threshing **126**
Storage **127**

Harvesting

Harvest within a week after 95% pods have turned yellow

Machine harvest

Hand picked

- Harvesting at the right time is critical to soybean seed quality and yield.
- When harvest is delayed soybean pods shatter, causing seed loss.
- Rain after pods have ripened will spoil seed quality.

Threshing

Sun or machine dried pods with less than 12% moisture

Beaten with a stick

Trampled by cattle

Threshed by machine

- Harvested pods should be well dried before threshing.
- Hand threshing is usually done by beating with a stick.
- For large-scale production, soybeans can be machine-threshed.

Storage

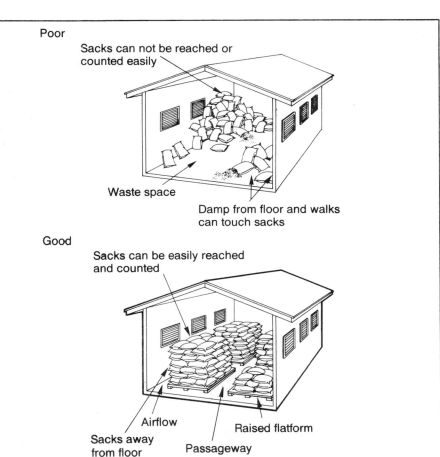

- Seed for storage should be dry, with no more than 12% moisture.
- The storage shed should be cool and dry.

Increasing yields and profits

Increasing yields and profits — yield components

Yield components **133**
Yield components — pods per plant **134**
Yield components — seeds per pod **135**
Yield components — seed weight **136**

Yield components

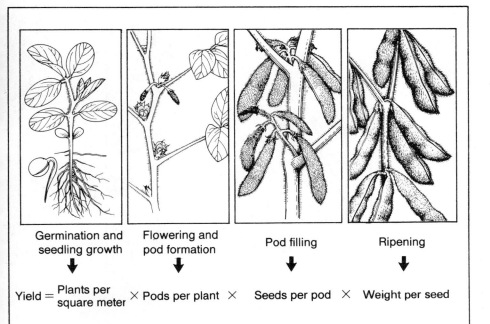

Germination and seedling growth → Flowering and pod formation → Pod filling → Ripening

$$\text{Yield} = \frac{\text{Plants per}}{\text{square meter}} \times \text{Pods per plant} \times \text{Seeds per pod} \times \text{Weight per seed}$$

- Every growth stage is important to total seed yield.
- Good management at all stages is needed for high yields. The growing conditions affect each stage of development.
- Reduction in any one of the yield components will reduce total yield.

Yield components — pods per plant

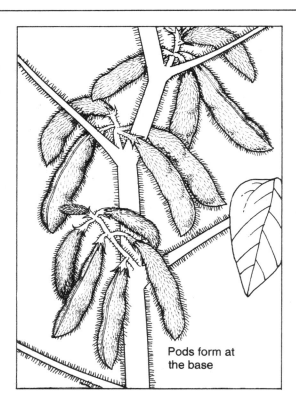

Pods form at the base

- The number of pods per plant is the most important yield component.
- About 40 percent of the flowers on a plant form pods. These can produce a good seed yield under favorable growing conditions.

Yield components — seeds per pod

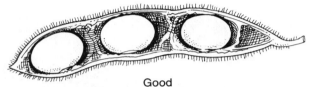

Good
Well-filled pods have no empty seeds

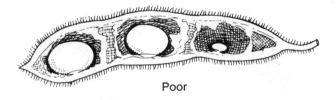

Poor

- The number of seeds per pod is determined at flowering, when the male pollen cell fertilizes the egg in the ovary.

Yield components — seed weight

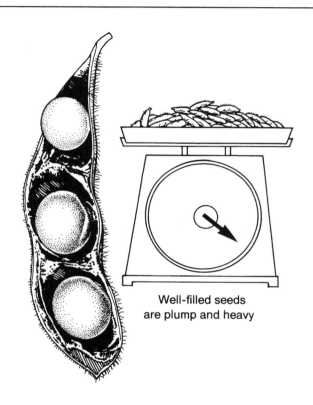

Well-filled seeds are plump and heavy

- Maximum seed size and weight depend on soybean variety.
- The weight of a seed is determined during seed filling.
- Drought or lack of nutrients at this stage will reduce the rate and length of time of seed filling.

Increasing yields and profits — production factors

Production factors   **139**
Planting improved varieties   **140**
Making the most of soil moisture   **141**
Using irrigation   **142**
Using fertilizer   **143**
Controlling pests and diseases   **144**

Production factors

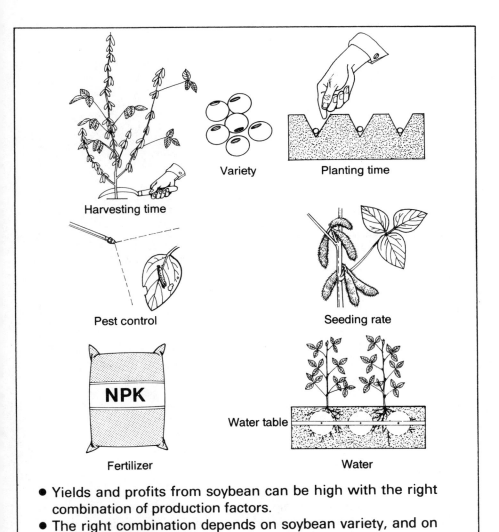

- Yields and profits from soybean can be high with the right combination of production factors.
- The right combination depends on soybean variety, and on season, location, and growing conditions.

Planting improved varieties

Plant varieties resistant to pests and diseases

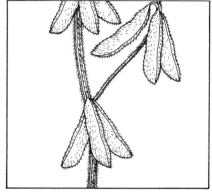

Plant high yielding varieties

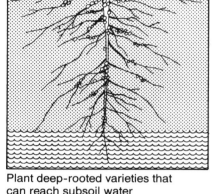

Plant deep-rooted varieties that can reach subsoil water

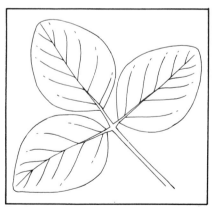

Resistant

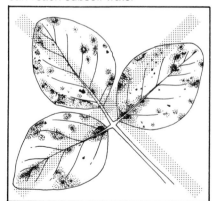

Susceptible

- Improved varieties give higher yields than traditional ones.
- Plant high-yielding varieties that are resistant to insects and diseases.

Making the most of soil moisture

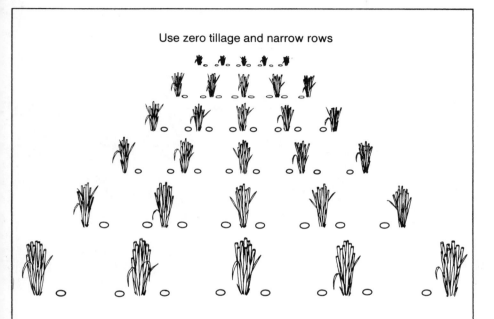

Use zero tillage and narrow rows

- In rainfed crops, making the most of soil moisture is the key to high yields.
- Plant soybean at once after the rice harvest. Or plant as a relay crop in standing rice 10 days before harvest.

Using irrigation

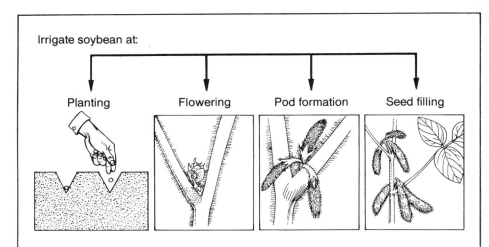

- Where water is available, irrigate soybean at planting and from flowering to seed filling.
- Good drainage is essential. Waterlogging will reduce yields.
- How much water to use depends on how much the soil can store.

Using fertilizer

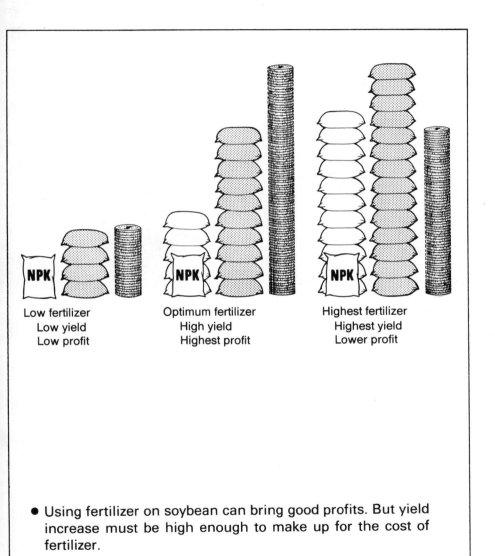

Low fertilizer
Low yield
Low profit

Optimum fertilizer
High yield
Highest profit

Highest fertilizer
Highest yield
Lower profit

- Using fertilizer on soybean can bring good profits. But yield increase must be high enough to make up for the cost of fertilizer.

Controlling pests and diseases

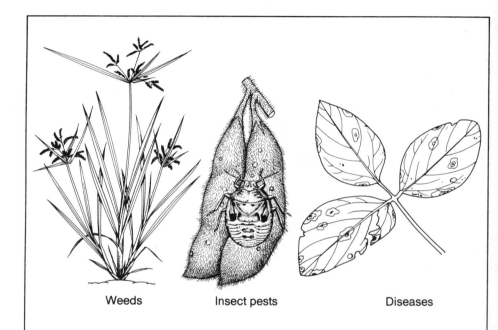

Weeds   Insect pests   Diseases

- Weeds, insect pests, and diseases can completely destroy a crop.
- Check these yield reducers early.

Yield reducers — weeds

Yield loss from weeds **147**
Weeds compete with soybean **148**
Weeds affect seedling growth **149**
Controlling weeds — by handweeding **150**
    Using cultural practices **151**
    By intercultivation **152**
    Using herbicides **153**
Common soybean weeds **154**
    Grasses **155**
    Grasses **156**
    Sedges **157**
    Sedges **158**
    Broadleaf weeds **159**
    Broadleaf weeds **160**
    Broadleaf weeds **161**

Yield loss from weeds

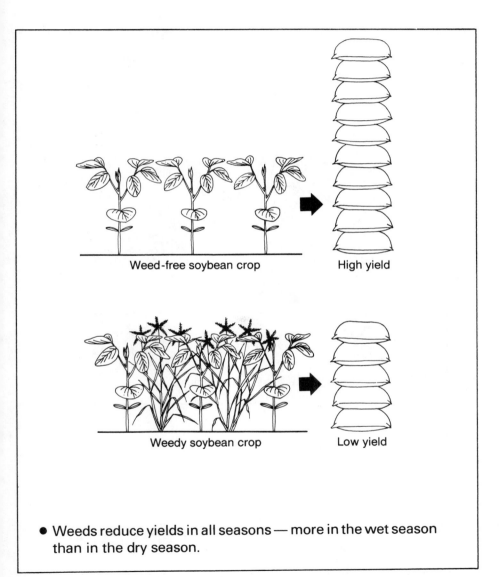

- Weeds reduce yields in all seasons — more in the wet season than in the dry season.

Weeds compete with soybean

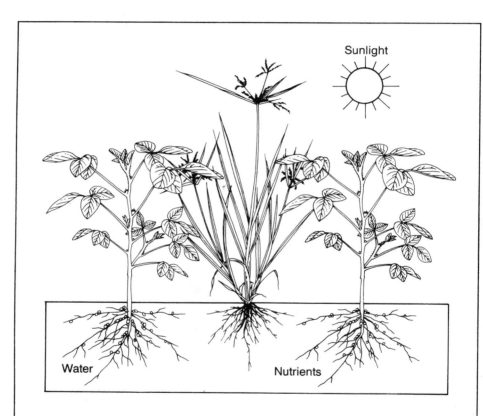

- Weeds compete with the soybean plant for sunlight, nutrients, and water.

Weeds affect seedling growth

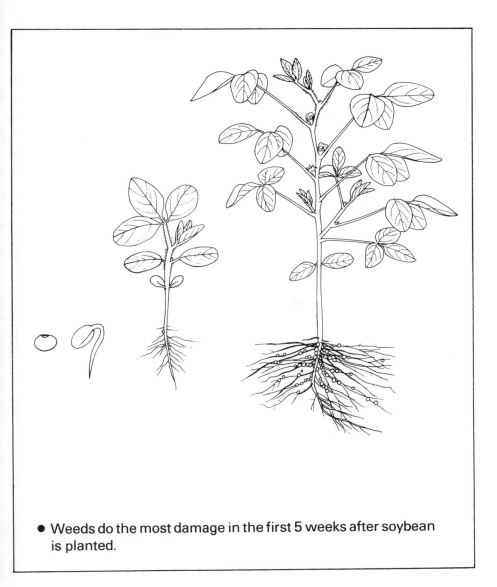

- Weeds do the most damage in the first 5 weeks after soybean is planted.

Controlling weeds — by handweeding

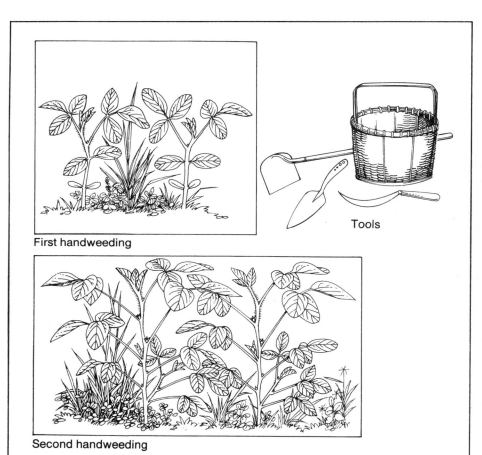

First handweeding

Tools

Second handweeding

- Weeds can be controlled by handweeding.
- Two handweedings are needed — one 2 weeks after planting and one at flowering.

Using cultural practices

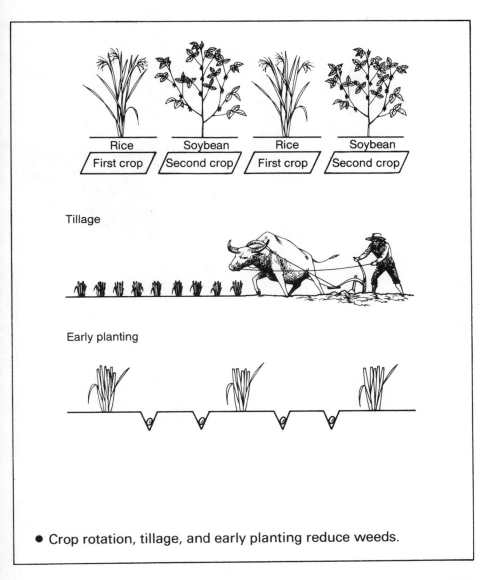

- Crop rotation, tillage, and early planting reduce weeds.

By intercultivation

Tractor

Animal-drawn

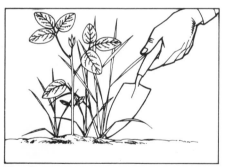
Hand

- Weeds can be controlled by intercultivation using a hand hoe or animal-drawn tools.
- For large-scale production, a tractor should be used.

Using herbicides

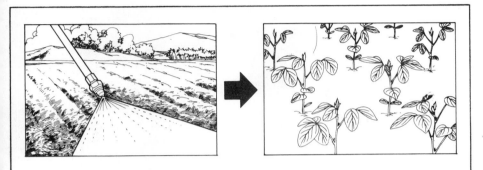

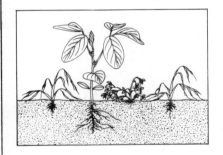

- Weeds can be controlled with chemical herbicides that kill the weeds but let soybean grow.
- Herbicides can be applied at planting to keep weeds from growing.

Common soybean weeds

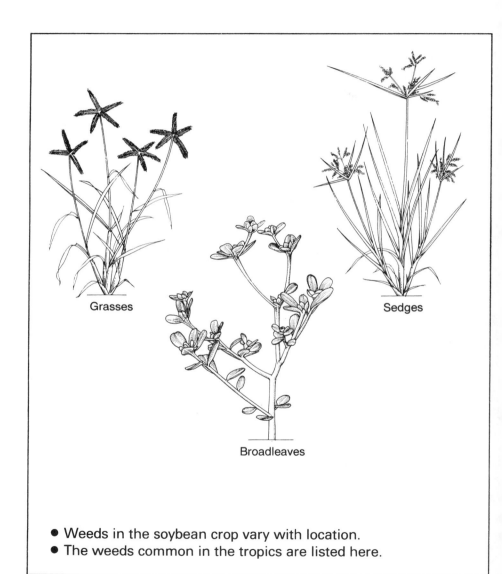

Grasses

Sedges

Broadleaves

- Weeds in the soybean crop vary with location.
- The weeds common in the tropics are listed here.

Grasses

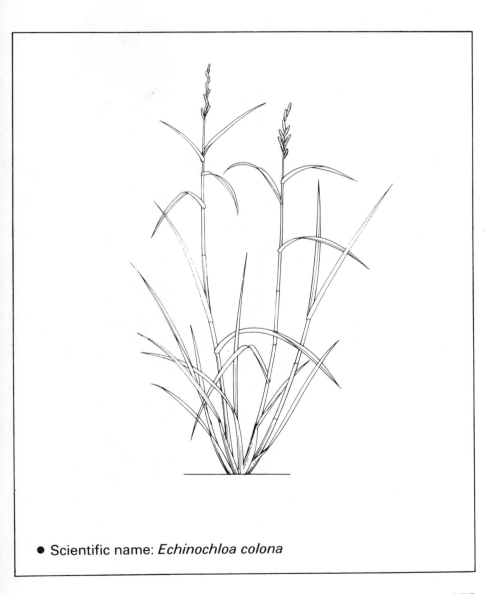

- Scientific name: *Echinochloa colona*

Grasses

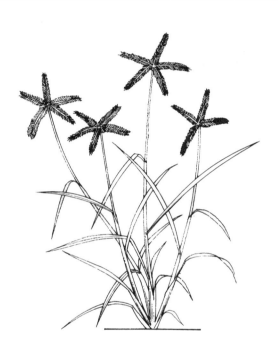

- Scientific name: *Dactyloctenium aegyptium*

Sedges

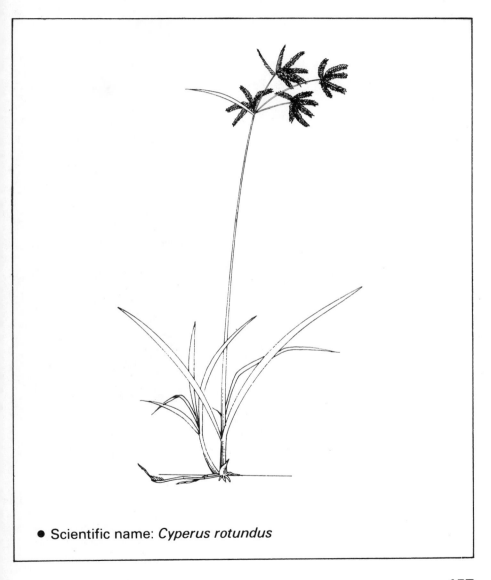

- Scientific name: *Cyperus rotundus*

Sedges

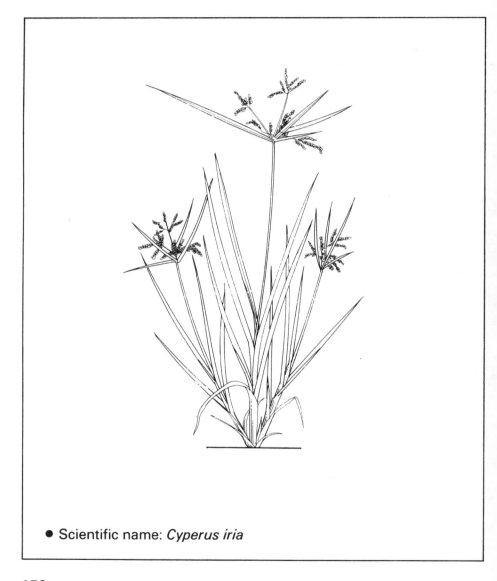

- Scientific name: *Cyperus iria*

Broadleaf weeds

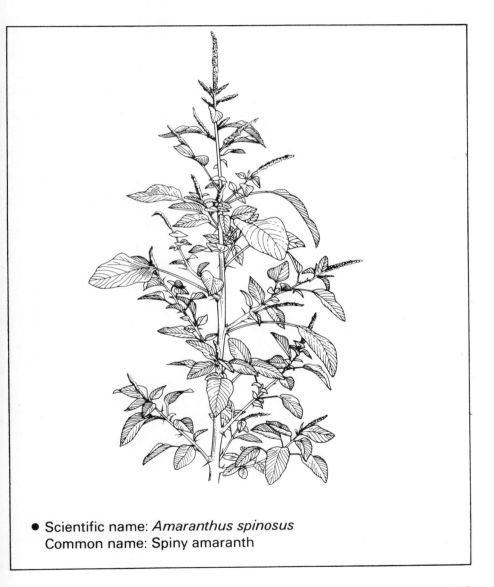

- Scientific name: *Amaranthus spinosus*
  Common name: Spiny amaranth

Broadleaf weeds

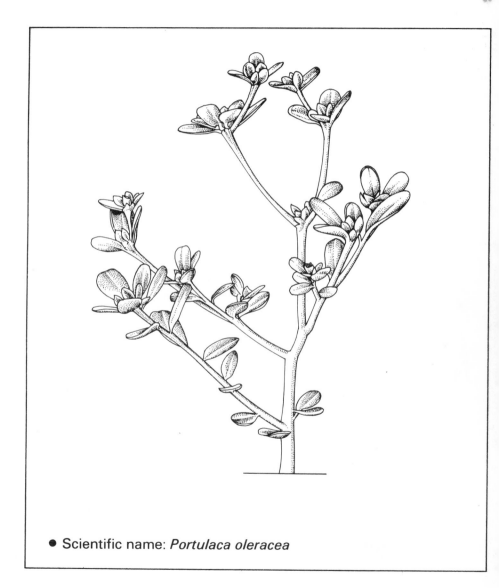

- Scientific name: *Portulaca oleracea*

Broadleaf weeds

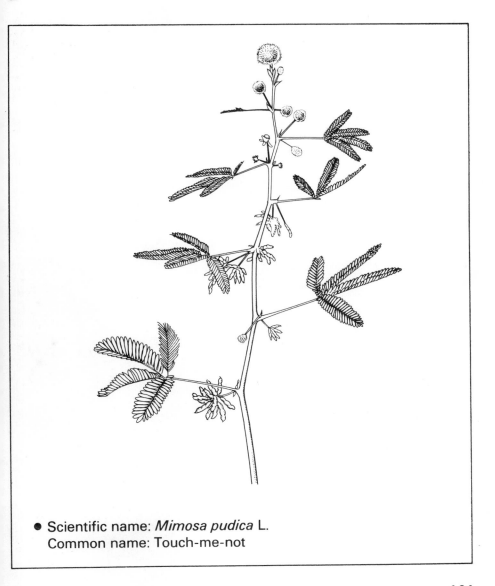

- Scientific name: *Mimosa pudica* L.
  Common name: Touch-me-not

Yield reducers — insect pests

Yield loss to insect pests **165**
Controlling pests — planting resistant varieties **166**
   Using cultural practices **167**
   Using insecticides **168**
   Combining pest control methods **169**
Common insect pests of soybean in the tropics — at seedling stage **170**
   At preflowering stage **171**
   At preflowering stage **172**
   Preflowering to pod formation **173**
   At pod development stage **174**
   At pod development stage **175**

Yield loss to insect pests

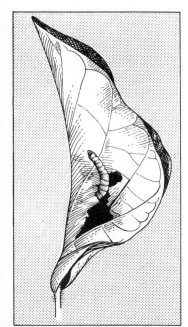

Leafminers and leaffolders reduce yields

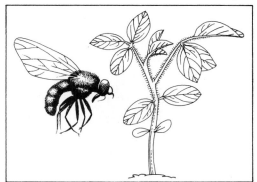

Beanflies stunt or kill seedlings

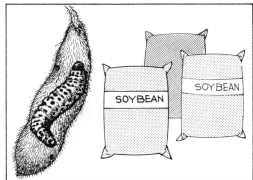

Podborers can cause as much as 80% yield loss

- Insect pests attack soybean at all stages of growth, from emergence to pod ripening.
- The most damaging pests vary with location and season.
- Yield loss depends on the growth stage at which the crop is attacked.

Controlling pests — planting resistant varieties

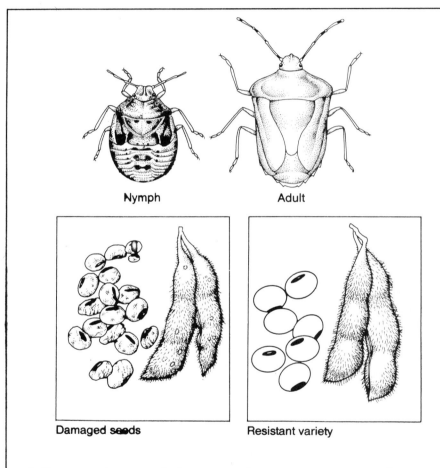

Nymph　　　Adult

Damaged seeds　　　Resistant variety

- Some soybean varieties are resistant to one or more insect pests.
- Planting resistant varieties is a low-cost way of controlling insect damage.

Using cultural practices

- Some insects can be controlled by cultural practices.

Using insecticides

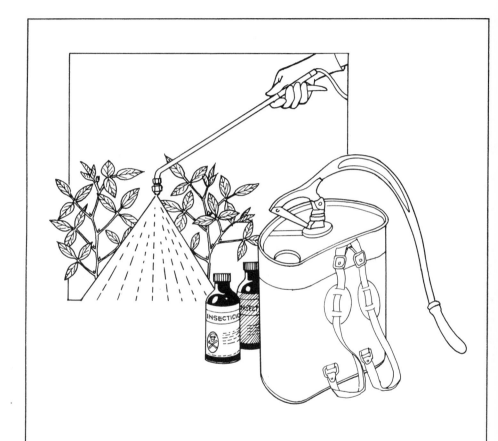

- Insecticides can be used to control pests, as needed.
- Strictly follow directions for use — time, method, and amount to be applied.

Combining pest control methods

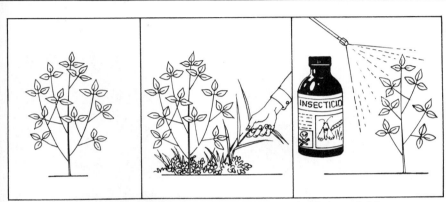

Plant resistant varieties   Cultural practices   Apply insecticides

- Several pest control methods can be combined:
    - planting resistant varieties
    - using proper cultural practices
    - applying the right insecticide at the right time.

Common insect pests of soybean in the tropics — at seedling stage

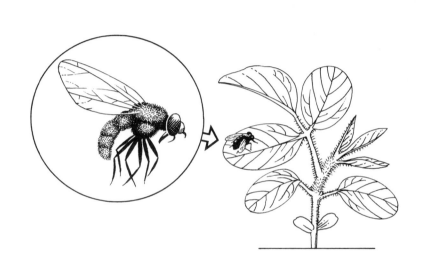

- Scientific name: *Melanagromyza sojae*
  Melanagromyza phaseoli
- Damage: Adults lay eggs on soybean leaves. Larvae tunnel through petioles and main stem. Seedlings are stunted or killed.
- Control: Use insecticide.

At preflowering stage

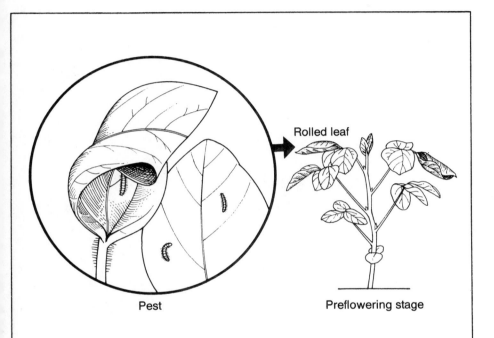

Pest   Preflowering stage

- Scientific name: *Hydylepta lamprosema*
- Damage: Larvae roll up leaves and feed inside. Leaves look silvery.
- Control: Use insecticide.

At preflowering stage

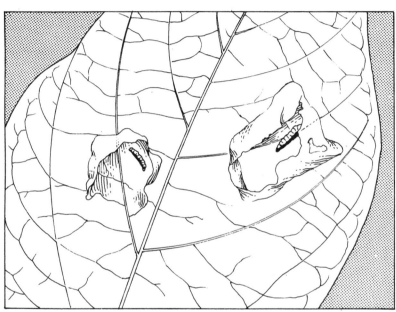

Leafminer *Stomopteryx subsecievella* (Zeller)

- Scientific name: *Stomopteryx subsecievella* (Zeller)
- Damage: The reddish caterpillars eat green portion of leaves; only a thin, silvery membrane is left. Leaves may drop off.
- Control: Use insecticide.

Preflowering to pod formation

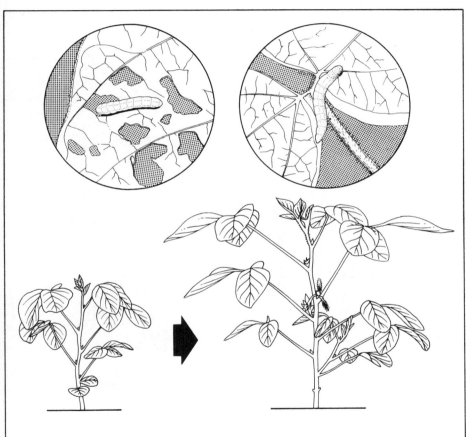

- Scientific name:
  1. *Diacrisia obliqua*
  2. *Spodoptera littoralis*
  3. *Heliothis armigera*
- Damage: Caterpillars feed on plant leaves and stems.
- Control: Use insecticide.

At pod development stage

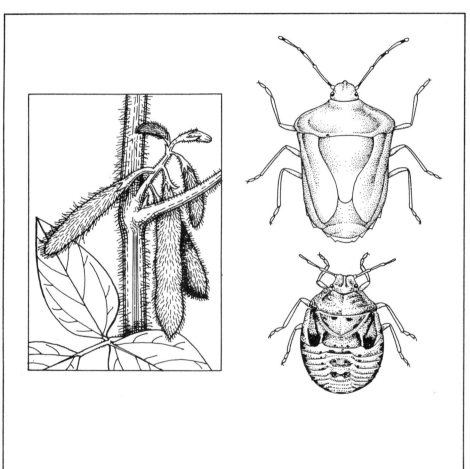

- Scientific name: *Nezara viridula*
- Damage: Young (nymphs) and adult stinkbugs suck juices out of pods and seeds.
- Control: Use insecticide.

At pod development stage

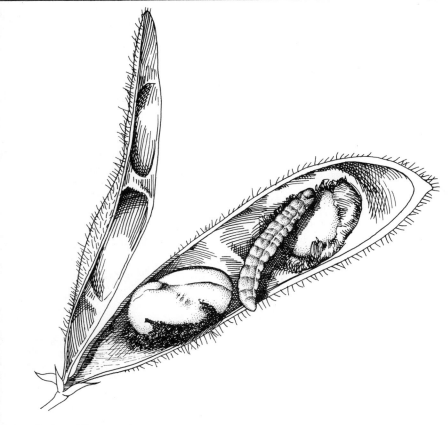

- Scientific name:
  1. *Etiella zinckenella*
  2. *Maruca testulalis*
- Damage: Larvae feed on developing seeds and cause heavy yield loss.
- Control: Use insecticide.

Yield reducers — diseases

Yield loss to diseases **179**
Controlling diseases — planting résistant varieties **180**
    Using cultural practices **181**
    Using chemicals **182**
Soybean diseases common in the tropics —
    Pythium seedling rot **183**
    Fusarium root rot **184**
    Rhizoctonia root rot **185**
    Phytophthora root rot **186**
    Charcoal rot **187**
    Anthracnose **188**
    Soybean rust **189**
    Purple seed stain **190**
    Bacterial pustule **191**
    Soybean mosaic **192**
    Bud blight **193**

Yield loss to diseases

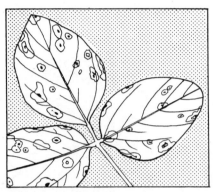
Leaf diseases

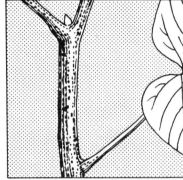
Stem diseases

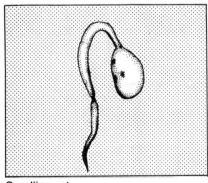
Seedling rot

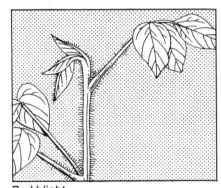
Bud blight

- Many diseases attack soybean and can severely reduce yields.
- Diseases and their severity vary with location and season. The ones common in the tropics are listed here.

Controlling diseases — planting resistant varieties

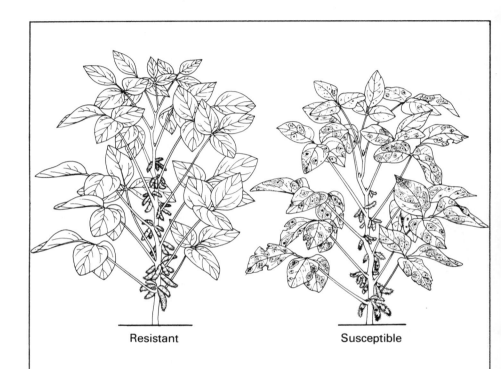

Resistant   Susceptible

- Some soybean varieties are resistant to one or more diseases.
- Planting resistant varieties is a low-cost way of preventing disease.

Using cultural practices

Deep plowing

Crop rotation

First crop | Second crop | First crop | Second crop

Intercropping

Corn　　Soybean　　Corn　　Soybean　　Corn　　Soybean

- Use cultural practices such as deep plowing, crop rotation, and intercropping to control diseases.
- Destroy crop residue that may shelter and spread disease.

Using chemicals

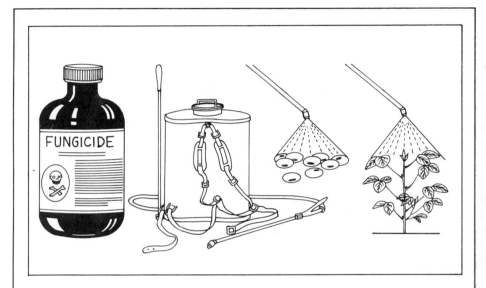

- Chemicals effectively control some diseases.
- Fungicides are especially useful in checking fungal diseases that attack seedlings and leaves.

Soybean diseases common in the tropics — Pythium seedling rot

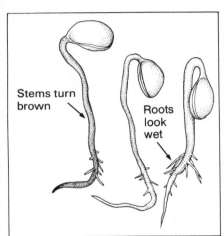

Caused by two kinds of soil fungus

Crop stand destroyed

- Scientific name: *Pythium ultimum* and *Pythium deburyanum*
- Symptoms: Roots look wet. Seedling turns brown.
- Control: Plant good quality, fresh seed. Treat seed with fungicide before planting.

Fusarium root rot

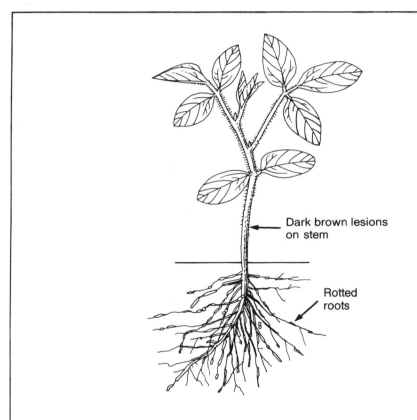

- Scientific name: *Fusarium oxysporum*
- Symptoms: Attacks in wet weather — heavy rain or flooded conditions. Seedling roots rot away; stems develop dark brown patches.
- Control: Plant good quality seed, treated with fungicide.

Rhizoctonia root rot

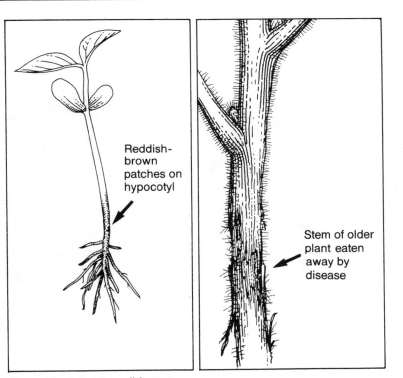

Caused by common soil fungus
Fungicide is not very useful against this disease

- Scientific name: *Rhizoctonia solani*
- Symptoms: Brown or reddish brown patches on lower stem and seedling hypocotyl.
- Control: Ridge soil around base of plants to reduce damage.

Phytophthora root rot

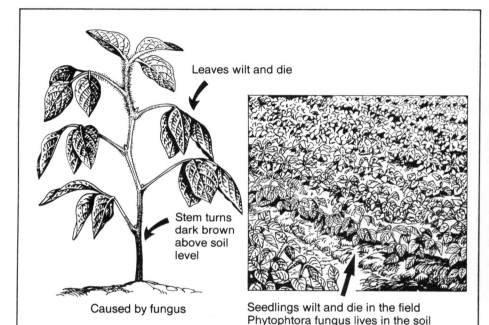

Leaves wilt and die

Stem turns dark brown above soil level

Caused by fungus

Seedlings wilt and die in the field
Phytophtora fungus lives in the soil from one season to the next

- Scientific name: *Phytophthora megasperma* var. *sojae*
- Symptoms: Stem just above soil line turns dark brown. Plants wilt and die. Common in low-lying, poorly drained areas and heavy clay soils.
- Control: Plant resistant varieties. Improve soil drainage.

Charcoal rot

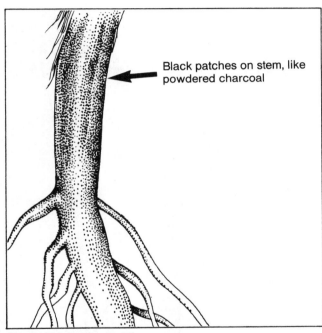

Caused by fungus that lives in dry soil

- Scientific name: *Macrophomina phaseolina*
- Symptoms: Lower stem shows black patches like powdered charcoal. Common in hot, dry weather and dry soil.
- Control: Crop rotation. Do not grow soybean in the same field in successive years.

Anthracnose

Dark brown lesions covered by dark structure

- Scientific names: *Colletrotrichum dermatium* var *truncata* and *Glomerella glycines*
- Symptoms: Anthracnose infects young seedlings and older plants. Dark brown patches appear on stem.
- Control: Crop rotation. Plant good quality, disease-free seed, treated with fungicide.

Soybean rust

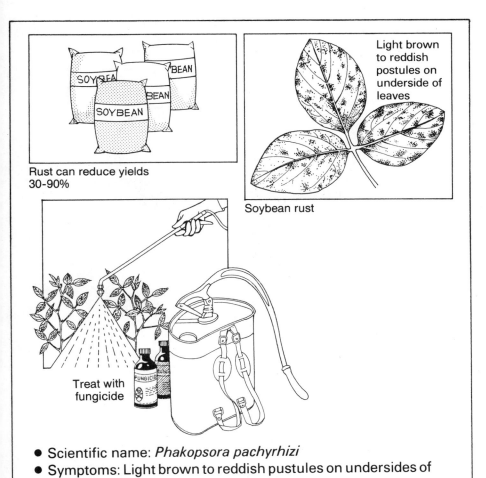

- Scientific name: *Phakopsora pachyrhizi*
- Symptoms: Light brown to reddish pustules on undersides of leaves. Leaves may drop off.
- Control: Use tolerant varieties that will yield even with rust attack. Treat with fungicide. No variety has been found to be free from this disease.

Purple seed stain

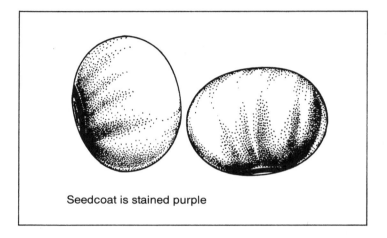
Seedcoat is stained purple

- Scientific name: *Cercospora kikuchii*
- Symptoms: Pale to dark purple staining of the seed. Infected seeds can produce diseased seedlings. Spreads to stem and leaves.
- Control: Treat seed with fungicide before planting.

Bacterial pustule

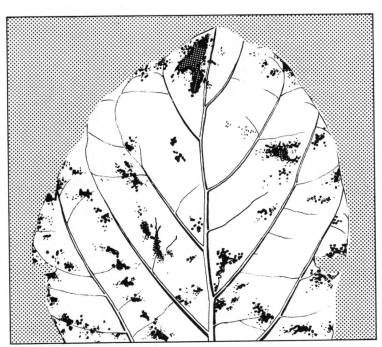

Brown spots with yellow edges on undersurface of leaf

- Scientific name: *Xanthomonas phaseoli* var *sojensis*
- Symptoms: Spots with brown center and yellow outer ring on under surface of leaf. Common in warm, wet weather.
- Control: Follow crop rotation. Plant resistant varieties.

Soybean mosaic

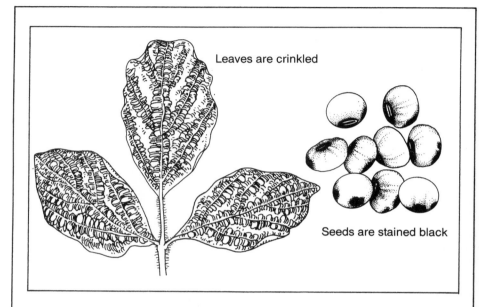

- Caused by soybean mosaic virus, which is seed-borne or carried by aphids from infected to healthy plants.
- Symptoms: Crinkled leaves, black-stained seeds.
- Control: Plant resistant varieties. Pull out and destroy infected plants from field. Plant new crops in disease-free fields.

Bud blight

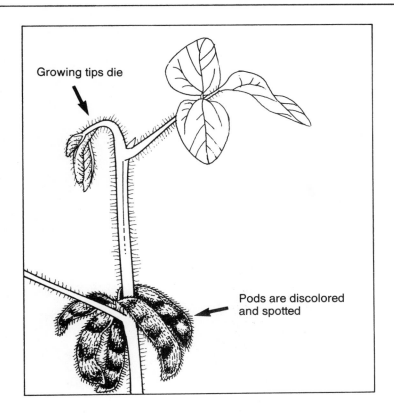

- Caused by tobacco ringspot virus.
- Symptoms: The top bud and shoot turn brown. Plant is stunted and remains green after normal plants have matured.
- Control: Pull infected plants from seed production fields. Do not plant soybean next to another legume crop.

Soybean in other cropping systems

Soybean in other cropping systems — sequence cropping

Soybean in other cropping systems   **199**
Sequence cropping — soybean before maize   **200**
   Soybean before sorghum   **201**
   Soybean before cotton   **202**
   Soybean before wheat   **203**

Soybean in other cropping systems

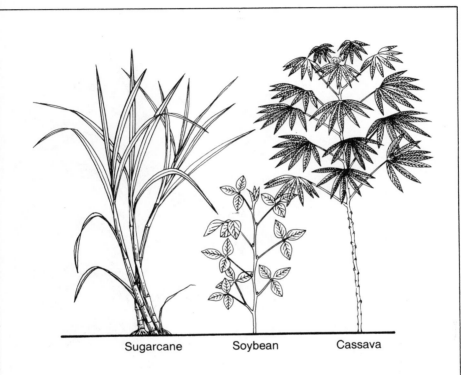

Sugarcane　Soybean　Cassava

- Soybean can also be grown with crops other than rice — with other cereals, sugarcane, cotton, or cassava.
- Short-duration varieties are especially suited to sequence cropping in upland areas.
- Soybean can be intercropped, strip-cropped, or grown in the spaces between plantation crops.

Sequence cropping — soybean before maize

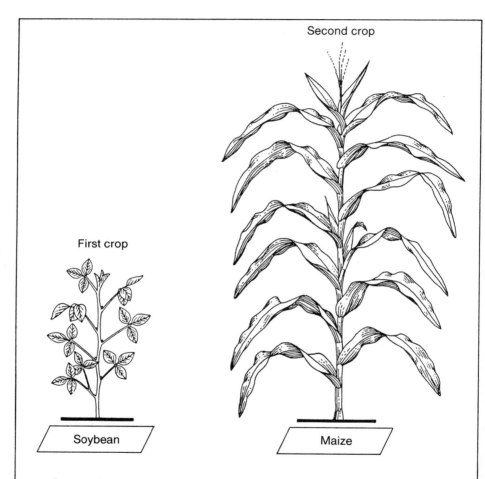

- Short-duration soybean is planted at the start of the rains in May. Maize is planted after the soybean harvest.
- The soybean-maize sequence is a more sustainable cropping system than continuous cereal cropping.

Soybean before sorghum

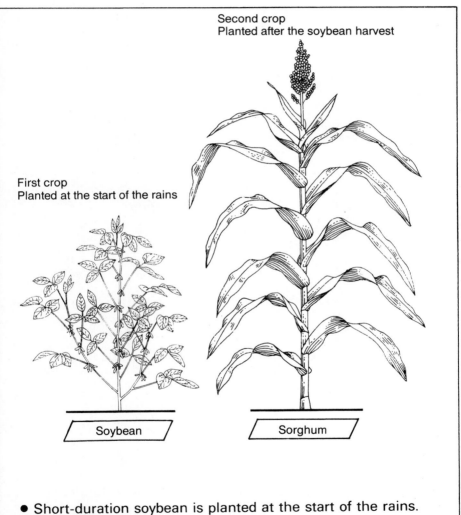

- Short-duration soybean is planted at the start of the rains. Sorghum is planted after the soybean harvest.

Soybean before cotton

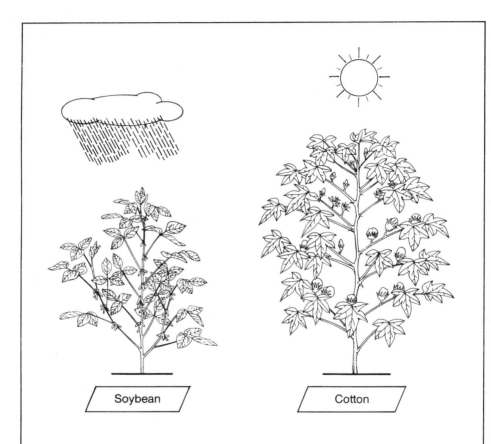

- Short-duration soybean can be grown during the early part of the rainy season.
- Cotton is planted at the end of the rains and harvested late in the dry season.
- This crop sequence gives good profits.

Soybean before wheat

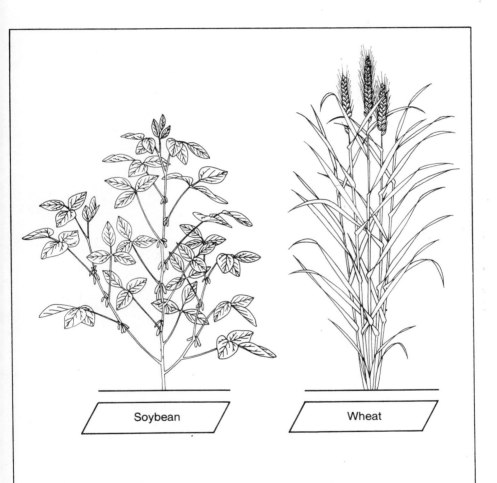

- In the cooler tropics where winter wheat can be grown, soybean can be planted as a first crop during the rainy season.
- Wheat is planted in November, and harvested in March-April.

Soybean in other cropping systems — intercropping

Intercropping — maize and soybean **207**
    Sorghum and soybean **208**
    Sugarcane and soybean **209**
    Cassava and soybean **210**
    Plantation crops and soybean **211**

Intercropping — maize and soybean

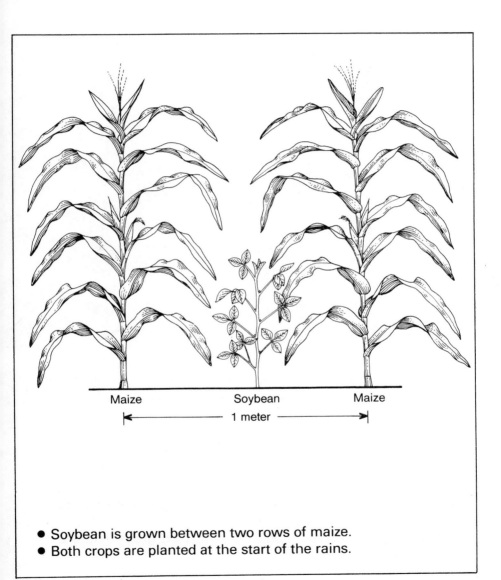

- Soybean is grown between two rows of maize.
- Both crops are planted at the start of the rains.

Sorghum and soybean

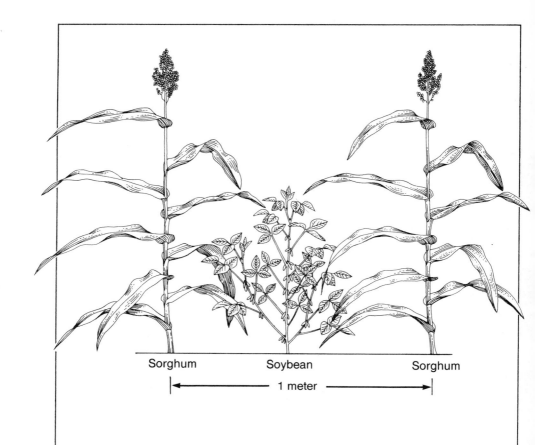

Sorghum — Soybean — Sorghum
|← 1 meter →|

- Soybean is grown between two rows of sorghum. Both crops are planted at the start of the rains.
- This cropping system produces more food grain from the same land.

Sugarcane and soybean

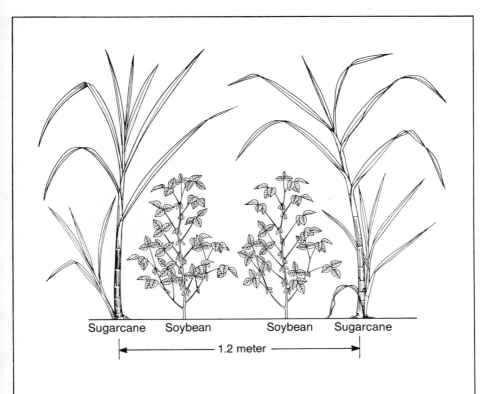

- Soybean is grown in paired rows between two rows of sugarcane.
- Both crops give good profits.

Cassava and soybean

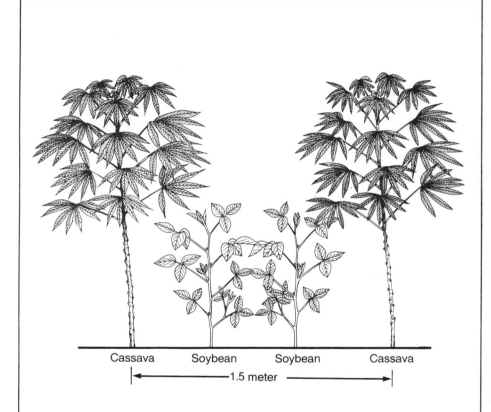

- Soybean is planted in paired rows between two rows of cassava.
- Soybean makes a nutritious supplement that adds protein to the starchy cassava diet.

Plantation crops and soybean

Banana and soybean

Rubber and soybean

Oil palm and soybean

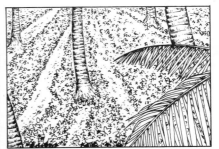
Coconut and soybean

- Soybean is planted in the vacant spaces of plantation crops such as coconut, oil palm, banana, and rubber.
- This makes full use of the land area and gives added income.

Soybean in other cropping systems — strip-cropping

Strip-cropping maize and soybean **215**
Strip-cropping sorghum and soybean **216**

Strip-cropping maize and soybean

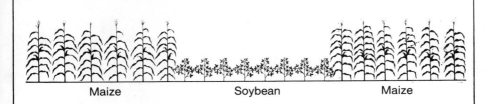

- Maize and soybean are planted in strips of six to eight rows each.
- In the next season, the two crops are rotated. Maize is planted in the place of soybean and soybean is planted in the place of maize.

Strip-cropping sorghum and soybean

First season

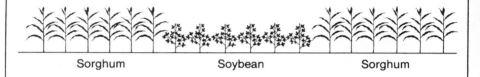

Sorghum   Soybean   Sorghum

Second season

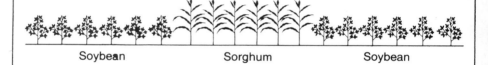

Soybean   Sorghum   Soybean

- Sorghum and soybean are planted in strips of six to eight rows each.
- In the next season, the crops should be rotated on the two strips.